SHINY OBJECTS

1/12

SHINY
OBJECTS

WHY WE SPEND MONEY WE DON'T HAVE
IN SEARCH OF HAPPINESS WE CAN'T BUY

JAMES A. ROBERTS

HarperOne
An Imprint of HarperCollins*Publishers*

HarperOne

HarperCollins books may be purchased for educational, business, or sales promotional use. For information, please write: Special Markets Department, HarperCollins Publishers, 10 East 53rd Street, New York, NY 10022.

HarperCollins website: http://www.harpercollins.com
HarperCollins®, ■ ®, and HarperOne™ are trademarks of HarperCollins Publishers.

FIRST EDITION

Designed by Janet M. Evans

Library of Congress Cataloging-in-Publication Data
Roberts, James A. (James Andrew)
    Shiny objects : why we spend money we don't have in search of happiness we can't buy / James A. Roberts.—1st ed.
        p. cm.
    Includes bibliographical references.
    ISBN 978–0–06–209360–8
    1. Consumption (Economics)—United States. 2. American Dream.
3. Materialism—United States. I. Title.
    HC110.C6R626   2011
    339.4'70973—dc22   2010005086

11 12 13 14 15   RRD(H)   10 9 8 7 6 5 4 3 2 1

I dedicate this book to my wife Julie and
daughters Chloé and Camille—
you have taught me what's really important
in life and through you I know what it is
to love and be loved.

CONTENTS

ACKNOWLEDGMENTS

To Julie, my lovely bride of 25 years I owe much. You have shown me unconditional love, support and a willingness to sacrifice for the good of the family since the beginning. Your honesty and sense of right are my guiding lights—testament to your parents, Jack and Jeanne Wilkes. Your bravery in the face of personal adversity has been inspiring. I will love you to the end.

And, with the birth of my daughters Chloé and Camille, I now know what it is to love someone more than you love yourself. My biggest wish is that these two young ladies are proud of who I have become. To be called "Dad" is the highest compliment I can receive.

I would be remiss if I didn't thank my mother, Carol Lassila, and my step-father Dennis. Mom always believed in me and provided the common sense advice only a mother can give. And, thanks to my step-father Dennis who introduced me to the wonderful world of academia which made this book possible. Through basketball, my father, Jack Roberts, showed me that anything was possible through hard work and determination. My older brother John has been a role-model for how to be a true professional and my younger brother Tom is someone who I have always wanted to be proud of me. My younger sisters Jackie and Rachel were wonderful additions to the family.

Writing this book has been a long time in the making. Growing up our family were recyclers, mindful of our impact on the environment and us boys were expected to work for anything more than the basics. It was here that we learned the value of money and the seeds of this book were born. My work in the banking industry and as a stockbroker opened my eyes to the important role money played in peoples' lives. My PhD dissertation addressed the topics of ecologically and socially conscious consumer behavior. This helped me to see how our behavior as consumers can impact others. *Shiny Objects* is the culmination of all this and 20 plus years of studying how our behavior as consumers can affect not only others but ourselves and the world around us.

I owe big thanks to my agent, Mollie Glick of Foundry Literary + Media. She took a chance and saw the potential for a project that others may have overlooked. Her nurturing approach to being an agent is much appreciated. Mollie took the time to help prepare a proposal that would catch the attention of an astute editor.

Cue Roger Freet of HarperOne. Roger saw merit in my proposal and played an important role in shaping *Shiny Objects*. Christina Bailly, Roger's assistant at HarperOne, provided much needed enthusiasm and assistance throughout the entire process. My production team of Carolyn Allison Holland and Terri Leonard did a super job in editing and playing traffic cop. Julie Burton, publicist at HarperOne, has been a tireless promoter for the cause.

I also offer a special thanks to Barbara Wiedman of Baylor University. She is the one-person secretarial pool in the Hankamer School of Business and did her typing of my nearly unreadable handwriting with a smile on her face, quick turnaround and a wonderful spirit—thank you Barbara. Co-authors Chris Manolis of Xavier University, Jeff Tanner of Baylor University, Steve Pirog of Seton Hall, Carol Gwin of Pepperdine University, and Carlos Ruy Martinez of Monterrey Tech University (ITESM) have made doing research both enlightening and enjoyable. Luminaries in the field of materialism research that have blazed a trail for this book include Tim Kasser, Marsha Richins, Russell Belk, Aric Rindfleisch, James Burroughs and others too numerous to mention. Graduate assistants, and future stars, John Beeson and Thanh Nguyen also provided much needed research assistance. And, last but certainly not least, I owe a deep debt of gratitude to Baylor University and the Marketing department who have given me the freedom to pursue topics of my choosing and have shown me and the world that faith and learning can co-exist very nicely.

SHINY OBJECTS

SHINY OBJECTS

The chief value of money lies in the fact that one lives
in a world in which it is overestimated.

—H. L. MENCKEN

Shiny object (shī´nē ŏb´-jĕkt´¹): anything that distracts the easily amused.

A "dog's life" never sounded so good. Forget about sleeping outside—42 percent of dogs now sleep with their owners and dine on organically grown meat, vegan snacks, and other gourmet treats. Many even get presents on their birthdays. Americans currently spend $48 billion a year on their pets. That's double the amount spent ten years ago and is more than the gross domestic product (GDP) of all but sixty-four countries. Spending on our furry friends is expected to top $58 billion in the next few years.

As a loving pet owner, you can splurge $535 on a dog ramp by Puppy Stairs to help your best friend make the ascent to your bed, $30 on an ounce of puppy perfume, and $225 on a trench coat for the family pooch. And let's not forget about doggie slippers, bikinis, and $500 Chanel pearls for those big nights out. What about a $270 Furrari bed for the little guy? Pet owners are also spending big on drugs to fight depression and separation anxiety in pets, as well as on psychological counseling, high-tech medical procedures, various cosmetic procedures, and end-of-life care. Plastic surgeons offer nose jobs, face lifts, breast reductions, braces, and tummy tucks for man's best friend. Nearly $10 billion is spent annually in the United States on veterinary services alone. Americans spend an additional $10 billion on over-the-counter drugs and supplies.

All doubt as to whether pet pampering is out of control ends with Neuticles, a patented testicular implant that fetches nearly one thousand dollars for a set of two. After pets have been neutered, Neuticles allows owners to "restore their pets to anatomical preciseness," and "preserve both their natural look and their self-esteem" according to inventor Gregg A. Miller who has sold more than 240,000 pairs of these little gems. Prosthetic testicles for your canine companion—only in America.[2]

PET PARAPHERNALIA IS ONLY THE TIP OF THE ICEBERG. WE ARE A NATION IN love with shiny objects. Our homes, our cars, our offices, our purses, and that storage unit we hate to admit to are all overflowing with our precious belongings. Whether your personal weakness is shoes, cars, jewelry, cigars, or any other possession (vintage posters, books, and watches are my downfall), we Americans love our stuff.

When it comes to spending money, are you more of a tightwad or a spendthrift? You'll have a chance to measure for yourself in chapter 5. Given that we are a nation of consumers, you might be surprised to learn that the majority of Americans would be classified as tightwads. With a high percentage of people living from paycheck to paycheck, how can consumerism be so rampant? It all boils down to how we pay for our purchases and the "pain of paying" associated with each payment method— it's not that tightwads don't want to spend money, they just don't want to feel like they're spending money. We are a nation addicted to plastic. Using credit cards greatly reduces the pain associated with paying for our purchases—so much so, in fact, that credit cards have earned the nickname "spending facilitators" by those of us who do research in this area. When we use credit cards, we make quicker purchase decisions, are more likely to buy, and are willing to pay more.

But can credit cards make us fat?[3] The answer to this question is an unqualified yes. When we use credit cards instead of cash at fast-food restaurants, we spend anywhere from 60 to 100 percent more. The average bill at McDonald's, for example, increased from $4.50 to $7.00 when customers started using credit cards instead of cash.[4] I call this the "supersize effect" of credit cards. If credit cards can expand your waistline or fatten your thighs, imagine what they can do to your household finances.

As a professor at Baylor University, I have spent over twenty years conducting research with thousands of consumers from all walks of life on the related areas of materialism, credit card use, and compulsive buying. Why, in a land of plenty, do Americans want more? And why is more never enough? Given that most Americans would readily admit that money and material possessions are not going to make us happy, why do we continue to act as if they will? This book is the culmination of my efforts and those of other researchers to answer such questions. And though consumers are inscrutable, we have begun to unlock some of the mysteries behind materialism and its impact on our lives.

As this book details, our obsession with possessions has a significant impact on our well-being. When asked what we really care about and what we consider to be the most life-giving elements of our existence, the vast majority of people respond in terms of the lasting value we place on our relationships with family and friends. And yet our consumer behavior contradicts such professed values. Our real habits—the time and resources we devote to accumulating more stuff—tell a different story. As the old saying goes, if you want to know what someone really cares about, look at that person's bank account.

It is my hope that reading this book will give you the time, space, and motivation to examine your day-to-day behavior in a way that our hectic lives rarely allow. Some of the studies and statistics I'll share may surprise you. Some may sound like they're describing someone else. But they all speak to one undeniable truth: as consumers, we're not who we think we are. It's time to bridge the gap between what we say and what we do. It's time to recommit ourselves to the kind of pursuits that are the true source of our well-being: spending time with loved ones, reaching our full potential as human beings, and participating actively in our world. No small task, but one well worth the effort: our happiness lies in the balance.

LIVING IN A MATERIAL WORLD

Our current consumer culture is best understood as an environment in which the majority of consumers avidly desire, pursue, consume, and display goods and services that are valued for non-utilitarian reasons such as status, envy, provocation, and pleasure-seeking.[5] Whether you reside in a major metropolitan city or a rural outpost of North America, you are part of the worldwide consumer culture. To ignore the importance of material possessions in our lives would be equivalent to ignoring that we are born of mothers and fathers.

The emergence of a worldwide consumer culture has potentially severe consequences for everyone. As you will soon see, even if you don't practice or espouse materialistic ideals, you are affected by others' pursuit of them. A good example might be the ghost of a recent Christmas past for retailing giant Walmart.

Surely a man the size of Walmart worker Jdimytai Damour could control the expected Black Friday shopping crowds. At six feet five inches and 270 pounds, he was a force to reckon with. In fact, he was chosen to work the front entrance to the Walmart store at the Green Acres Mall in Valley Stream, New York, precisely because of his hulking frame. But, alas, he was no match for the crowd of 2,000 Walmart shoppers eagerly awaiting the 5:00 a.m. store opening. A few minutes before store opening the throng could no longer be held back. The sliding glass doors that separated the would-be shoppers from the myriad of holiday bargains ("door busters" takes on a whole new meaning) bowed from the bodies pressed against them. Six to ten workers attempted to no avail to push back, but they were fighting a losing battle. In an instant, the glass doors shattered and the frenzied mob surged into the store in search of the heavily discounted "doorbusters" available in limited quantities for a short period of time. Tragically, Mr. Damour was thrown to the floor and trampled to death (the official cause of death being asphyxiation related to his trampling) in the stampede that streamed over him in pursuit of bargains on big-screen TVs, electronics, clothing, and a myriad of other consumer goodies. One shopper, Kimberly Cribbs of Queens, said that the crowd acted like "savages." And the shoppers' bad behavior didn't end with the trampling of Mr. Damour. When the shoppers were informed that the store would need to be cleared because of the death of an employee, many continued to shop, yelling that they had been waiting in line since the day before. Many had to be escorted from the store.[6]

The *Oxford English Dictionary* defines materialism, in that word's common usage, as "devotion to material needs and desires, to the neglect of spiritual matters; a way of life, opinion, or tendency based entirely upon material interests." I think of materialism as a mind-set, an interest in getting and spending, the worship of things, the overriding importance that someone attaches to worldly possessions. For a consumer who has fully embraced shiny objects, possessions take center stage and are considered to be the primary source of all happiness. Money and material possessions are seen as an end in themselves, rather than as a means to an end. Materialism is the cornerstone of our modern consumer culture.

We Americans attempt to find happiness and satisfaction through the acquisition of possessions, which typically assume a central role in our life. Of course, not all Americans are equally materialistic, but on average we are a materialistic lot. Those of us who are highly materialistic (let's use the term "materialists") believe that expanded levels of consumption will increase the amount of pleasure we achieve in life. Research, however, paints a bleak picture for happiness through acquisition, consistently showing that those of us who live materialistic lifestyles are less happy with our lives than less materialistic people are. On average, U.S. consumers are no happier than less profligate consumers around the world. I will have more to say about the materialism–happiness relationship in chapter 4.

Materialists tend to judge their own and others' success by the number and quality of accumulated possessions. The primary value of possessions, for diehard materialists, is their ability to confer status and project a desired self-image. Materialists view themselves and others as successful only to the extent that they possess products that project a desired image. How successful can my colleague be, they wonder, driving a car like *that*, or living in *that* neighborhood? Judging others by what they possess is a deeply ingrained part of our collective psyche.

In our rush, rush world, a common way we tell others who we are (or would like to be) is through our use and display of material possessions. He drives a Mercedes, so he must be a captain of industry. A truck and he must be a cowboy or at least a rugged individualist. A Hummer, and . . . I'm not sure what that says about the driver. This tendency to define ourselves by the products we consume results in what researchers call the "extended self"; in other words, our possessions become an *extension* of who we are. I amend the label to the *"over*extended self" when referring to consumers who have fully embraced the shiny-objects ethos. Research has found that highly materialistic people value their possessions for their ability to conjure up a desired social image, whereas their less materialistic brethren value their stuff for the pleasure and comfort it provides. Furthermore, as you've probably noticed with your more pretentious acquaintances, materialistic people are more likely than less materialistic people to mention an item's financial value when describing why it's important to them: "That cost me nearly $30,000!"

Not only do a person's materialistic values affect how he or she relates to possessions, but they also affect how that person spends money. Compared to less materialistic people, materialists believe that they require more money to satisfy their "needs" and are more likely to spend money on themselves and friends and donate less to charities.

The "Great Recession" of 2008/2009 and the continuing economic malaise have the average folks on Main Street on edge regarding their financial future. With high unemployment, mortgages being forfeited, and credit being tightened, it's likely that Americans will be spending less this year— and that's a terrifying thought for citizens who have been taught that shopping is a patriotic act. We live in a country where we are repeatedly told that happiness can be purchased at the mall, online, or from a catalog,[7] so the idea of scaling back on our purchases is frightening. But I've got some good news to share: happiness is *not* positively correlated with consumption.

Between 1972 and 2010, the standard of living in America increased dramatically. When we produce more, we consume more[8], and within that time period, our national output per capita increased by 96 percent. But

U.S. GDP and Happiness, 1972–2010:
What's Wrong with This Picture?

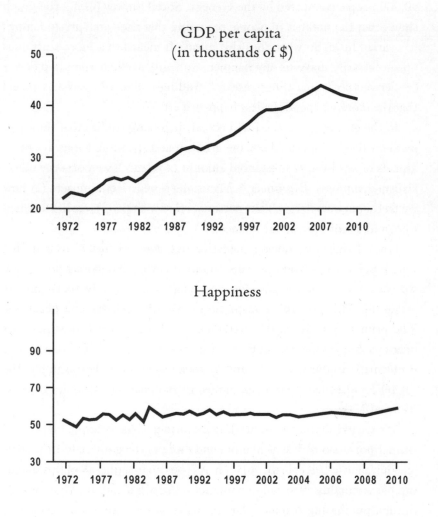

GDP for earlier years has been adjusted to current dollars. Happiness data was taken from the General Social Survey (GSS)[9] of over 50,000 people and represents the percentage of people who responded "pretty happy" to the question: "Taken all together, how would you say things are these days—would you say that you are very happy, pretty happy, or not too happy?"

while our standard of living has improved over the past thirty to thirty-five years, our happiness has not. In fact, as the graph on the previous page depicts very clearly, our happiness has flatlined. Based on surveys of over 50,000 people conducted by the General Social Survey (GSS), the graph shows that the number of people reporting that they are "pretty happy" has varied little. So what does this mean? It means that more stuff does not necessarily make us any happier. And this particular study paints a "rosier picture" than other research findings—alternate surveys reveal that the more we spend the less happy we are.

In the chapters to come I will reveal surprising studies that showcase just how deeply ingrained our materialism and spending habits are in all aspects of our lives. A substantial amount of research supports the materialism–happiness disconnect.[10] Materialism negatively impacts (1) how we feel about ourselves, (2) our personal relationships, (3) our life satisfaction, and, (4) of course, our finances.

One of the more obvious negative outcomes of materialism is that fourth point: the havoc it wreaks on our finances. No matter how much we consume, we never get closer to happiness; we only speed up the treadmill. This process of adaptation has both positives and negatives. The primary positive is that we adjust quickly to most bad things that befall us; however, we adapt to good things equally well, and that can be problematic to our happiness and financial well-being. For example, the (huge) 2,500-square-foot house almost instantaneously becomes the new "normal."

We are like drug addicts, needing a continuous fix of newer, bigger, and shinier possessions. We need more and more of the good stuff to achieve an equivalent (albeit ephemeral) high. As you can imagine, all of this spending has a ruinous effect on our finances. We are a nation of compulsive buyers, purchasing products far in excess of our needs and resources. Outside of our basic need for food, shelter, and clothing, everything else we purchase is discretionary. Given that nearly three-quarters of all U.S. families live paycheck to paycheck,[11] it is obvious that many of us struggle with money issues. Ask yourself the following questions:

1. Do you have an emergency fund of at least $2,500 saved for that proverbial rainy day?

2. Do you have six months' worth of living expenses in the bank in case you lose your job or become ill?

3. Have you put away money for your kids' college fund(s)?

4. Are you regularly making investments into a retirement account, whether a Roth or a regular IRA?

If you answered no to any of the above questions, it's time to take a careful look at the role money and material possessions play in your life.

THE HIGH COST OF MONEY

Are you like me? At the start of every New Year I pull out a piece of paper and write down my goals for the coming year. You know the typical . . .

Lose ten pounds.

Exercise more.

Read more books.

Save more money.

Watch less television.

Spend more time with the family.

Cut back on various bad habits *(feel free to fill in the blanks below).*

    a. Bad habit: _____

    b. Bad habit: _____

    c. Bad habit: _____

I bet none of us have goals like the following:

Live paycheck to paycheck.

Spend more money than I can afford.

Max out any available credit cards or lines of credit.

Pay off old credit cards with new credit cards.

Buy a car I can't afford.

Fall behind on mortgage payments.

Borrow money at usurious rates—or worse, from friends and family.

Spend less time with the kids and more time at work.

Take out payday loans.

Confuse needs and wants.

Although we might not have goals like these, we sure act like we do. In fact, these unspoken aims may be the only goals we achieve from year to year.

Goals are important to us, whether they are written down or are simply embodied in the values we hold. A highly materialistic lifestyle, one based on goals like the second set above, comes at a high cost, as we will see in chapter 7. In our dogged pursuit of material possessions, we leapfrog the cornerstones of psychological well-being: self-acceptance, intimate relationships, and community involvement.

FALLOUT FOR RELATIONSHIPS

Never after a devastating natural disaster like a tornado or flood have I seen someone interviewed on TV say, with relief, "Well, it got the kids, but at least it left the flat-screen TV." Yet our behavior as consumers seems to suggest such skewed priorities. We act as if we prefer money over people.

In a fascinating series of experiments, researchers Kathleen Vohs, Nicole Mead, and Miranda Goode found that highly materialistic people opt for less social intimacy, particularly if there's a financial advantage in doing so. They also discovered that merely providing subtle reminders about money (in the form of a computer screensaver with money floating across the screen, for example) changed people's willingness to help others. In the first of three experiments that looked into that second issue—helping others—subjects in the experiment played the board game Monopoly, after which the participants moved on to a new task. Before the new task began, however, participants assigned to what the researchers called the "high-money condition" were given $4,000 of Monopoly money, while people in the "low-money condition" were given $200; all were told the money was to be used later. Participants in the control group played Monopoly but received no money afterward. After each of the participants finished playing Monopoly they were led across the laboratory by the experimenter for the purported purpose of performing a task in another room. As the participants crossed the room, however, a confederate (someone employed by the researchers) walked by and dropped a batch of pencils in front of them. Amazingly, participants who had been given the larger sums of Monopoly money were less helpful in picking up pencils than either the low-money or control groups.

In a second experiment in the series, high-money-condition participants doing a puzzle exercise were less likely to help a confused peer (actually another confederate of the study) ostensibly having trouble understanding the instructions for a task she was attempting to complete. Participants in the low-money condition spent 120 percent more time helping the confused student than those in the high-money condition.

A third experiment investigated whether reminding participants of money would have an impact on their generosity. Prior to the manipulation (reminded of money or not), participants were paid eight quarters ($2). Participants were then either subtly reminded of money or not before being given the opportunity to make a donation to the university student fund. As you might have guessed, participants who had been reminded of money donated a smaller percentage of their money than those who had not been reminded of money (39 percent compared to 67 percent).

Most startling about this series of experiments is the fact that the mere mention of money was found to lead people to act in a more selfish, non-generous manner. It is evident from the three experiments described above that even subtle reminders of money have a significant impact on our helping behavior. From a relationship standpoint, being reminded of money makes people less sensitive to the needs of others. As the authors conclude, "even subtle reminders of money elicit big changes in human behavior."[12]

MADONNA GOT IT RIGHT

Are Americans becoming more materialistic? To answer that question we need look no further than the spending habits of the typical American consumer. Our standard of living has soared since the end of World War II. In fact, if you make a graph of American lives since then, every line that has anything to do with money and material possessions shoots straight upward. Our national zeitgeist is one of happiness through consumption.

And consume we have. Fifty years ago the average U.S. home was approximately 1,200 square feet. By the mid 1970s, the average home size had grown to 1,645 square feet. By 2008, the average home was approximately 2,500 square feet.[13] All this while families are getting smaller, not bigger. Forget the two-car garage; any "respectable" new home today comes standard with a three-car garage. The same story holds for the number and size of cars we drive. We want our transportation shiny, fast, and big. The latest cell phones and other electronic gadgets are considered de rigueur. One of my regrets in all of this consuming is that I didn't invest in the storage unit business at the start of this consumer orgy.

And it's not just the parents, either. In fact, contemporary teens and preteens are the most brand-conscious and materialistic generation to date. American children feel that the clothes they wear and the brands they use tell the world who they are and define their social status. Parents, teachers, and social observers are becoming increasingly concerned over the rising tide of materialism in children and adolescents. In a recent national survey, 90 percent of adults say that children are too preoccupied with buying and consuming, with nearly 80 percent agreeing that limits need to be placed on advertising to children. Over $1 billion per year is spent on media advertising to children, with an additional $10 billion spent on promotions, public relations efforts, and purchasing designed to attract children.[14]

A recent survey of about a quarter million college freshmen provides additional evidence of our growing obsession with possessions.[15] The survey, conducted by UCLA on an annual basis and released most recently in January of 2010, found that 77 percent of all freshmen surveyed thought it was important to be "very well-off financially." Compare this with the 62.5 percent of freshmen who agreed to the same in 1980 and 42 percent in 1966. This increasing materialism in college students is consistent with my own research. With each subsequent generation (Baby Boomers, Gen Xers, Gen Yers, and Millennials), I have found higher incidences of credit card abuse and compulsive buying. It appears that we are becoming increasingly dysfunctional in regard to how each generation spends money. For many young adults, the money involved in credit card transactions is so abstract as to be almost unreal. Debt has stopped being a four-letter word for many young people. A 2009 MetLife study of the American Dream found that 81 percent of Gen Xers admitted to not having an adequate financial safety net despite the bad economy.[16]

Additional evidence of our preoccupation with consuming can be found in the financial fallout from all our purchases. The two charts below show just how far our spending has gotten out of hand. Although consumer bankruptcies ebb and flow, the trend line in the first chart reveals a steady increase since the early 1980s, with a high-water mark of over 2 million people filing for bankruptcy in 2005.[17] Even with a tightening of the consumer bankruptcy laws in that year, the current economic crisis will likely see a further increase in bankruptcy filings. Credit card debt is a significant portion of debt for individuals filing for bankruptcy. As can be seen

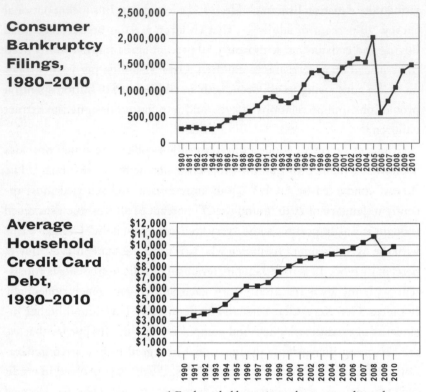

Consumer
Bankruptcy
Filings,
1980–2010

Average
Household
Credit Card
Debt,
1990–2010

* For households owning at least one credit card.

in the second chart, Americans are in love with their credit cards. Average credit card debt has risen steadily since the early 1990s—hitting its peak in 2008 when the average American household possessed thirteen credit or charge cards and carried nearly $11,000 in credit card balances.[18]

Further proof of our current consumption binge is literally piling up in the trash produced by the American consumer culture. We toss out 140 million cell phones each year, use 1 million brown-paper bags each hour (yes, hour), expend 1 million plastic cups on commercial airplanes every six hours, and run through 2 million petroleum-based plastic bottles every five minutes.[19] Some time ago, retail analyst Victor Lebow described the situation like this:

> Our enormously productive economy . . . demands that we make consumption a way of life, that we convert the buying and use of goods into rituals, that we seek our spiritual satisfaction, our ego satisfaction, in consumption. . . . [W]e need things consumed, burned up, replaced, and discarded at an every-increasing rate.[20]

The singer Madonna was right: we *are* "living in a material world." According to our government, all that materialistic consumption is a patriotic act, and according to popular culture, retail therapy is a great cure for the blues. But if, as we are beginning to see, materialism doesn't make us happy and instead runs us into debt and buries us in disposability, why *are* we this way? More importantly, how can we change?

WHY THE OBSESSION WITH POSSESSIONS?

If I were to ask you, "What's the first thing that comes to your mind when I say 9/11?" I bet few of you would say, "Shopping." Yet shopping played a central role in the aftermath of the 9/11 tragedy. Shortly after the horrific event, President George W. Bush encouraged Americans to go shopping. "We can't let the terrorists achieve the objective of frightening our nation to the point where we don't conduct business, where people don't shop," exhorted the president.[21] And we listened. Americans went on a post-9/11 shopping spree, buying everything from canned goods and comfort foods to such big-ticket purchases as RVs, homes, and expensive cars.

Shopping as a means of coping with the anxiety and sense of doom associated with a devastating event such as 9/11 may, at first blush, seem a bit strange. There is, however, quite a simple explanation for such behavior: terror management theory (TMT). According to TMT, humans are unique in their awareness of death and the anxiety that it creates. This awareness motivates people to enrich their lives with meaning and bolster their sense of worth by adopting important cultural values. In a consumer culture, says TMT, people turn to material possessions to cope with not just tragic events but also the mundane anxieties and stressors of everyday life.[22] We'll look at that phenomenon in chapter 9.

Even without the trigger of terrors, big and small, people have a fundamental tendency to adopt cultural values and behaviors via a process called "internalization." As we live and work in today's consumer culture, where economic indicators are the primary barometers of well-being and money and material possessions are seen as ends in themselves, we take on those values. In fact, there's a whole vast marketing machine designed specifically to encourage us to do so. Experts claim we are exposed to as many as 3,000 marketing messages daily.

A trusty weapon in the marketing arsenal is product placement, whereby companies pay movie and television producers to display products in their shows. Next time you watch your favorite TV show or go to a movie, see how many brands you can pick out. You'll be amazed. The recent hit movie *Marley & Me* had twenty-seven product placements—a few more than the average of twenty-four for most popular movies. The movie *Paul Blart: Mall Cop* had an astounding forty-nine product placements. Iron Man 2, the 2010 superhero flick starring Robert Downey Jr., contained a league-leading 64 product placements.[23] But even these impressive numbers are put to shame by the film industry's closest relative—TV. It would be a gross understatement to say that TV has embraced product placements. *The Biggest Loser* TV show had 4,000 product placements in its first sixteen episodes of 2008. That's an eye-popping 250 product placements per show. The wildly popular *American Idol* may be the prime example for product placements, with 4,151 over its first thirty-eight episodes of 2008.[24] Research is clear on the impact of TV-watching on materialism: simply put, people who watch TV are more materialistic.

Even if you're relatively savvy about the sales pitches and product placements that marketers use to get you to buy their products, you may be shocked by their latest and greatest technique: getting everyday people to extol their products' virtues. Beware, MySpace and Facebook denizens. These days, blogs that appear, on the surface, to be one person's opinion about life are often paid advertising supported by brand marketers. Marketers have even co-opted word of mouth. That guy next to you in the bar singing the praises of Absolut vodka may actually be a paid spokesman for the vodka manufacturer. The same for the Office Depot devotee at work: there's a good chance she gets free products to encourage her to spread the good word.

Sunday morning is not exempt from this growing trend. Many churches and church leaders encourage a materialistic lifestyle. The so-called prosperity gospel, also called the "name it and claim it" theology, preaches that authentic religious beliefs (and regular financial contributions to church) will result in material prosperity. This particular gospel, discussed in chapter 10, contends that financial prosperity and success in your private and professional lives is evidence of God's favor. Religious leaders who preach the prosperity gospel are growing in number and strength.

Another pathway to materialism, often overlooked, is that materialism may be part of our DNA. As humans, we may be programmed to acquire and consume more and more resources. This desire for more was helpful when we were still living in caves in a resource-strapped world, but it's not so well suited to the abundant world we now inhabit. Recent advances in neuroscience demonstrate that genetics play a big role in determining human behavior, and it may well be that biology determines how much self-control you're able to display in the face of the consumer culture. Is behavior modification impossible? No. However, science suggests that there may be limits.

BREAKING THE CHAINS OF MATERIALISM

Understanding the trade-offs of a materialistic lifestyle and recognizing the marketing machine behind our consumer culture are important first steps toward becoming less materialistic. The next step—developing your self-control—is a bit more difficult. Conjure up a mental image of a loud party next door when you're watching reruns at home alone. You know that the partygoers will have a hangover the next morning, but it's still hard not to join the party. Self-control is the ability to delay gratification, as we will see in chapter 12. You rely on self-control any time you place a priority on future outcomes, overlooking smaller, more immediate rewards. Self-control is what allows you to save money for the inevitable rainy day, your retirement, or your children's college education instead of spending it on short-term pleasures such as restaurant meals, unnecessary clothes, fancy cars, jewelry, or vacations. It's the skill set that allows you to pursue subtler, more long-term goals like close relationships with others, rather than going for the immediate gratification of a night in front of the TV. According to Sigmund Freud's theory of human development, the ability to delay gratification signifies the transition from an impulse-driven existence to a higher-order, future-oriented perspective. Freud saw the ability to delay gratification as the core of personality development and the foundation for any civilized society.[25] It's also the key to becoming less materialistic.

At the very heart of self-control is the ability to control our impulses. An impulse is a strong desire to perform a particular action—whether it's going to the mall, visiting eBay online, or browsing catalogs when you're bored. Impulsive behavior is based on a sudden whim or desire rather than on careful, thoughtful deliberation. People often blame "irresistible" forces or "uncontrollable" urges for their lack of self-control, but research does not support such a notion. There is no such thing as an irresistible impulse. Exclaiming "The devil made me do it!" may make you feel better in the moment, but it won't help you avoid the same behavior in the future.

The good news is that although the pull of impulses is strong, there are simple ways you can reduce their power. The single most effective way to improve your self-control is by creating an environment that sets the stage for behavioral change. Research has supported the axiom that "It is easier to avoid temptation than it is to resist it." Experiments have shown that the likelihood of a behavior is strongly influenced by the presence or absence of environmental cues associated with that behavior. For instance, you're more likely to buy things you don't need if you've got six catalogs sitting on your desk. In chapter 13 I provide twenty-five ways to "tweak" your environment to help control your spending and materialistic longings.

THE AMERICAN DREAM

The American Dream is a belief in the freedom that allows all citizens of the United States to pursue their goals in life through hard work and free choice. It's about opportunity—a chance to better oneself. These days it usually includes some discussion of a car in the driveway, a house with a white picket fence, and two and a half kids. Although the meaning of the American Dream has evolved since this country was founded, one component of the national dream seems to be fairly consistent: the quest for money and financial independence. To remove economic opportunity from the American Dream would be to remove its very core.

The traditional message of the American Dream was that through hard work, frugality, and sacrifice, anyone could achieve financial independence. Somehow we lost our way on the road to that dream. Presently, many Americans have replaced the traditional American Dream with a philosophy of "get rich quick." Dreams of easy money have replaced hard work, thrift, and self-sacrifice.

In fact, many social critics would argue that Americans are not embracing the American Dream as much as the American Daydream. The hard work and sacrifice that were part and parcel of the original American Dream have been replaced by wishful thinking about material success with no willingness to pay the dues necessary to create such wealth. A 2009 MetLife study found that 50 percent of Americans could survive only a month without a paycheck before going into debt. Twenty-eight percent of these same people said they couldn't last two weeks without a steady paycheck.[26] We want to enjoy the trappings of wealth and "look the part," but we don't like the sustained self-sacrifice component.

Instant wealth wasn't originally an important component of the American Dream. In fact, Benjamin Franklin's bon mots included "A penny saved is a penny earned." According to Mr. Franklin, the key to financial independence is industry: "Early to bed, early to rise, makes a man healthy, wealthy, and wise." And, unlike today's version of the American Dream, the primary goal of those early Americans was not extravagant wealth but simple economic independence—the opportunity for a better life for oneself and one's family.

The tragedy of allowing the American Dream to be defined solely in economic terms is that we have sold it short. Freedom from want is only one of the many freedoms offered under the American Dream. In addition to the freedom from want, the American Dream also offers freedom to speak our minds and the right to worship as we see fit. It is an audacious dream—a dream that *all* Americans, regardless of race, creed, or social class, are free to make a better life for themselves. Although somewhat battered and bruised, both maligned and venerated, and continually evolving (as documented in the next two chapters), that lofty dream is the United States' legacy to the world.

The evolution of the American Dream may be the second-greatest story ever told. It is an epic tale of a people searching for meaning. At first, it was enough to have the freedom to worship, speak one's mind, and have enough food to eat. Our Puritan founding fathers viewed work as a calling from God.[27] They believed that people could serve God through their daily toil. And thus the seeds of capitalism were sown. But it would take several hundred years and considerable tumult before those seeds grew into anything approaching the present consumer culture.

As our material standard of living continued to increase, Americans' desire for creature comforts grew with it. The Declaration of Independence guaranteed the pursuit of happiness as a right of all Americans. It was then up to individuals to define the specifics of that grandiose promise. Increasingly our focus turned to material possessions.

History is messy, but I have been able to identify a number of pivotal events in our nation's history that shaped the American Dream—some so lurid that you will cringe and shake your head when you read them. You can't tell millions of people that they can achieve the American Dream of owning a home easily—with no money down and an adjustable rate mortgage (ARM) or interest-only payments—without that audacity eventually biting you in the rear end. As columnist Thomas Friedman puts it, "If you jump off the top of an 80-story building, for 79 stories you can actually think you're flying. It's the sudden stop at the end that always gets you."[28] The next two chapters chronicle the events that transformed the American Dream from the asceticism of our founding fathers to the no-holds-barred, consumers-gone-wild mentality that in turn led to the recent mortgage crisis and the near collapse of the U.S. financial system.

CHASING THE AMERICAN DREAM

There are those who will say that the liberation of humanity, the freedom of man and mind is nothing but a dream. They are right. It is the American Dream.

—ARCHIBALD MACLEISH

I, like many of my fellow Baby Boomers, am the living embodiment of the American Dream. We are better off (variously defined) than our parents, who themselves were a rung higher on the socioeconomic ladder than their parents. My maternal grandfather was a hard-working truck driver and my grandmother ran the household. Their daughter (my mother), now seventy-two and a former schoolteacher, is a college graduate with a master's degree.

My paternal grandfather was a salesman with a stint in the buggy trade. He was a bit older than my grandmother, who took care of their two children and ironed anything and everything in sight. My dad was a high school and college basketball star who coached baseball and basketball and taught high school mathematics for over thirty years. Along the way he also earned his master's degree.

After graduating from college, I spent approximately four years in banking and sales before I decided to go back to school to earn my MBA. It was during my graduate work that I realized that I, too, would like to be a teacher. Following my father's advice to avoid teaching at the high school level—it's hard to support a family on a teacher's wages, he warned—I enrolled at the University of Nebraska, where I earned my PhD in marketing. I have taught classes in consumer behavior and advertising at the college level since 1991.

ALL THIS TO SAY THAT THE AMERICAN DREAM IS REAL. IT IS NOT SOME PIPE dream, but the guiding principle of a social order that offers opportunity to all of its citizens—a chance for men and women of all races and nationalities, from even the lowest rungs of the economic ladder, to attain their fullest stature. It is a dream that has been realized more fully in America than anywhere else in the world. You need look no further than your own family history to see the dream at work. The American Dream is our legacy to the world. The French may have given the world *la belle cuisine* and the Italians art, but the United States has held out the prospect of hope and opportunity to the millions who make it their home.

But now the American Dream is on life support—it's not dead but teetering on the brink of oblivion. Our banking and mortgage industries, crippled by avarice, have been propped up by billion-dollar bailouts. The U.S. auto industry was nearly destroyed, and it was changed forever in the upheaval. Millions of homeowners are in foreclosure or unable to make their mortgage payments; millions more have burdensome credit card debt and a nonexistent savings rate.

Not since the Great Depression have we seen such carnage. Why, you might ask? The simple answer is that many of us got lost somewhere along the road to the American Dream. How could something that started out with the noblest of intentions devolve into little more than a mad dash for material possessions?

GOD AND THE AMERICAN DREAM

The Protestant work ethic has been a core component of the American Dream since the founding of the United States. What started overseas in the sixteenth century formed the very basis of our capitalistic society. The Protestant Reformation in western Europe brought to the American shores a radical restructuring of the importance of work. The two sixteenth-century reformers responsible for that shift were Martin Luther and John Calvin. The Protestant Reformation addressed the underlying tension between the everyday exigencies of life and the more spiritual aspects of human existence.[1]

While previous cultures had seen little spiritual value in work and in many cases considered it demeaning, Martin Luther—breaking from the Catholic Church—viewed work in a very different light. He believed that God was served through one's work, that the trades were beneficial, and that work was the glue that held society together. A person's work, asserted Luther, was his calling, and it should not be challenged or changed. A key tenet of Luther's thinking was that all callings were of "equal spiritual dignity," placing manual labor on equal footing with other occupations. Luther's thinking fell short of what we might consider capitalism, however, because he disapproved of the buying and selling of products as an occupation. Luther argued that each person should earn a sufficient income to cover the basic needs of food, shelter, and clothing, but to amass money and material possessions was a sin.

It was the layering of John Calvin's theological doctrines on top of Luther's foundation that ultimately led to a markedly different perspective regarding work. Calvin was a French theologian known for his doctrine of predestination. That doctrine maintained that only the elect, those chosen by God, would inherit eternal life. All the rest of us were damned, and there was nothing we could do about it—God was unchanging. The sticking point was this: How did you know whether you were one of the lucky few? The simple answer was that you didn't. However, you could search for evidence of God's favor in your life. The best evidence was thought to be found in your daily actions and in the success you enjoyed in your daily endeavors. A slacker was no doubt one of the damned, but an active, devout, and hard-working person was surely (hopefully?) one of God's elect.

In Calvin's thinking, work was the will of god. As professor Roger Hill puts it, "It was the duty of men to serve as God's instruments here on earth," and that "men were not to lust after wealth, possessions, or easy living, but were to reinvest the profits of their labor into financing further ventures. . . . Selection of an occupation and pursuing it to achieve the greatest possible profit was considered by Calvinists to be a religious duty." Sanctifying the pursuit of unlimited profit, as Calvin did, asserts Hill, "was a radical departure from Christian beliefs during the Middle Ages."[2]

The great French social thinker Alexis de Tocqueville, in his 1835 book, *Democracy in America*, saw the real genius of America in the early nineteenth century as its "productive industry"—a bent toward productivity that had not yet, as author Steven Malanga puts it, started its "descent into lethal materialism."[3] This precarious balancing act, mused de Tocqueville, was maintained because Americans shared a common set of civic virtues that went beyond hard work alone and encompassed thrift, self-sacrifice, honesty, and self-reliance. All of those virtues, noted de Tocqueville, shared a religious component that enveloped American democracy and economy in a moral blanket.

It was about seventy years later that renowned German economist and sociologist Max Weber coined the phrase "Protestant ethic" to summarize the virtues identified earlier by de Tocqueville as the driving force behind America's dedication to work. It was the virtues of hard work, thrift, sacrifice, honesty, and work as a calling that Weber considered the sine qua non of a thriving capitalistic economy.[4] Weber wrote that capitalism emerged when the Protestant ethic led large numbers of people to join the secular workforce, undertaking various business ventures, and making profits to be invested back into the business. He considered the Protestant ethic to be the "force behind an unplanned and uncoordinated mass action that influenced the development of capitalism," and, in turn, the American Dream.[5]

CHARTER OF THE AMERICAN DREAM

The evolving religious perspective that grew out of the Protestant Reformation influenced this country while it was still a collection of colonies. It was a young nation's cry for independence from the British that next shaped the American way of life. Many consider our nation's Declaration of Independence to be the charter for the American Dream. The under-girding for that dream can be found in the opening salvo of its second paragraph, which reads: "We hold these truths to be self-evident, that all men are created equal, that they are endowed by their Creator with certain unalienable rights, that among these are life, liberty and the pursuit of happiness."

Most Americans have a vague recollection of most of this and many even remember those seven immortal words, "life, liberty and the pursuit of happiness." Later in the same paragraph, primary author Thomas Jefferson wrote that when government runs amok, "it is the right of the people to alter or to abolish it" and to form a new government whose purpose is to organize itself in such a way "as to them shall seem most likely to effect their safety and happiness." According to author and historian Jim Cullen, "the pursuit of happiness" is the "phrase that more than any other defines the American Dream, treating happiness as a concrete and realizable objective."[6]

The Declaration of Independence cannot be considered a highly original work. Not surprisingly, the desire to be happy had been a topic of much discussion for some time. Some years after the document's drafting, John Adams, himself a contributing author, stated that there was "not an idea in it, but what had been hackneyed in Congress two years before."[7] But in Jefferson's defense, it was not meant to be a document that grew forth from his own idealistic ambitions, but rather a distillation of the many heated debates and impassioned pleas that had occurred in Congress and its anterooms in the preceding years.

A good example of how Jefferson borrowed from others can be found in John Locke's 1689 *Second Treatise on Government*. At one point in this work, Locke advocates "life, liberty and the pursuit of property." It is likely that Jefferson used Locke's earlier work to help him capture the founding principles of an emerging nation. In our attempt to assess the current state of the American Dream, we see that Locke, with "property," may have come closer to capturing its core element than Jefferson, with the more ambiguous "happiness." Jefferson and his contemporaries were never really able to define the "pursuit of happiness." And that's just the point; they left it up to you and me to fill in the blanks.

Although defined differently from person to person, "the pursuit of happiness" has brought and continues to bring millions of immigrants to our shores in hope of a better life and endless opportunities. Such high-minded thinking emboldens the rest of us to persevere in tough economic times with the fervent belief of better days ahead. In fact, the Economic Mobility Project, a 2009 nationwide survey initiated by the Pew Charitable Trusts, found that although the current economic crisis is front and center in the minds of the American people, an undercurrent of optimism keeps bubbling to the surface.[8]

Of the thousands of Americans polled by Pew, 79 percent feel it's still possible for people to get ahead in the current economy, 72 percent believe they will be better off economically in the next ten years, and 74 percent feel that "being free to accomplish anything" (in other words, to pursue happiness) best describes the American Dream. Most noteworthy is the finding that the American Dream transcends racial boundaries. Little difference was found between how Caucasians, African Americans, and Hispanics define the American Dream, although African Americans are the most optimistic about their economic future (85 percent), compared to whites and Hispanics at 71 percent and 77 percent.

Despite such optimism, we all know there's no guarantee that we will ever achieve happiness. Simply the right to pursue happiness in its various forms is enough for most of us.

CALIFORNIA DREAMING

Steven Malanga claims that the Protestant ethic flourished for three centuries in America. He notes that "the breakup of the 300-year-old consensus on the work ethic began with the cultural protests of the 1960s, which questioned and discarded many traditional American virtues."[9] No doubt the 1960s were turbulent times, but cracks in the approval rating of the Protestant ethic appeared much earlier. The California Gold Rush of the late 1840s is a good example of how the key virtues of the Protestant ethic—hard work, sacrifice, and temperance—were slowly being replaced by the "get rich quick" mentality.

It was a winter's day in January of 1848 and James Marshall was busy building a sawmill for John Sutter alongside the American river located to the Northeast of what is now Sacramento, California. It was here that Marshall's gaze fixed on a shiny object on the bank of a creekbed near a location known to the local natives as Cull-u-mah (Coloma).

When Marshall shared his discovery with his employer at Sutter's Fort, some distance away, the two agreed to keep Marshall's discovery a secret for the moment. Sutter intended to complete his sawmill before the news spread. Confident that a run on gold was imminent, he hoped to capitalize on the resulting need for lumber. It wasn't long before word of their little secret started to spread, however. To prevent wholesale desertion when some of Sutter's other workers discovered gold, Sutter allowed prospecting on weekends and off-hours. Nonetheless, rumors of gold floated downriver to San Francisco.[10]

Sam Brannan, an enterprising merchant, caught wind of the rumors of gold and decided to create his own gold rush. What tipped him off? A customer from Coloma paid for his supplies in Brannan's Sutter's Fort dry goods store with gold dust. Brannan promptly purchased every available shovel and pickax, marking them up for resale, and procured a small vial of gold. To ignite the smoldering embers of gold fever, Brannan sloshed the muddy streets of San Francisco yelling, at the top of his lungs, "Gold! Gold! Gold in the Sierra foothills!" The race was on. Local residents responded enthusiastically, leaving the town nearly deserted in their search for gold. The cries of "Gold!" caused such a stir that the once thriving San Francisco harbor quickly became a ghost town. Entire crews struck by gold fever abandoned their vessels as they floated in the harbor. San Francisco's two newspapers shuttered their operations—their staffs decimated by gold fever. When one of them, the *Californian*, stopped production, it left behind a final editorial announcing, "The whole country, from San Francisco to Los Angeles, and from the sea shore to the base of the Sierra Nevada resounds with the sordid cry of 'Gold, Gold, Gold!' while the field is left half-planted, the house half-built, and everything neglected but the manufacture of shovels and pickaxes."[11]

When the *New York Herald* printed a story about the discovery in August of 1848, the Gold Rush officially became a stampede. In December of 1848, President James Polk validated the discovery when he said, "The accounts of the abundance of gold in that territory are of such extraordinary character as would scarcely command belief were they not corroborated by authentic reports of officers in the public service." Horace Greeley of the *New York Tribune* added further fuel to the flame when he wrote, "Fortune lies upon the surface of the earth as plentiful as the mud in our streets. We look for an addition within the next four years equal to at least one thousand million of dollars to the gold in circulation."[12] By early 1849, gold fever was rampant. Discussions about gold and the prospect of instant riches were the topic of conversation across the United States and around the globe. The backwater town of San Francisco had mushroomed to a sprawling metropolis with a population of about 100,000 by late 1849.[13]

As booms are wont to do, this one was followed by a bust. In a few short years, the easy pickings were gone, and so were many of the prospectors. But California's name and mythology were to be forever tied to the Gold Rush and what became known as the "California Dream." Even today, California is seen as a land of opportunity and new beginnings, where hard work and a little luck can lead to great wealth. Historian H. W. Brands notes that after the Gold Rush the California Dream spread to the rest of the United States and became part of the new American Dream. As Brands puts it,

> America was becoming more like California. . . . The old American Dream, the dream inherited from ten generations of ancestors, was the dream of the Puritans, of Benjamin Franklin's Poor Richard[,] . . . of men and women content to accumulate their modest fortunes a little at a time, year by year. The new dream was the dream of instant wealth won in a twinkling by audacity and good luck. . . . James Marshall's discovery electrified the country (and the world), holding forth the promise that wealth could be obtained overnight, that boldness and luck were at least as important as steadiness and frugality.[14]

A CHANGE IN HUMAN CHARACTER

The American Industrial Revolution, which began in the mid-1800s and continued well into the twentieth century, had a profound impact on Americans—not only because of the dizzying array of products it produced, but also because of how it transformed our priorities. Americans no longer sought the "good life"; rather, they sought a "life of goods." A shift to modernism was taking place across the globe. As Virginia Woolf has been quoted as saying, "On or about December 1910, human character changed."[15] Hyperbole aside, there is more than a little truth in that statement. The long-entrenched Protestant ethic of hard work, thrift, and self-discipline was quickly being replaced with a more hedonistic approach to life and an about-face on saving and spending. Life was to be enjoyed, and individual fulfillment was paramount. The "Gospel of Relaxation" was shared by business managers and the educated elite as well as factory workers and tradespeople.

This cultural shift was required to accommodate the ever-increasing supply of consumer goods rolling off conveyor belts across America. American output per worker increased 60 percent from 1870 to 1900.[16] While this increase in productivity formed the basis for the nascent consumer culture, that culture was being shaped by more than the simple abundance of goods. When I teach about this period in American business history, I refer to it as the sales era—a time characterized by the injunction, "Send out the truck full (of unsold products), and bring it back empty." We were now so good at making things that we needed to find ways to get rid of them. By 1900, nearly 1,200 catalogs took the latest in housewares and high fashion to even the remotest outposts. Opulent department stores first appeared in the last third of the nineteenth century, as did the less opulent but farther-reaching purveyor to the masses— Sears. The Sears catalog found its way into 3.8 million homes by 1908. Similarly, F. W. Woolworth's chain of variety stores brought less expensive versions of fancy department store products to 600 small towns by 1912.[17] Advertising was also growing, selling not only products but the consumer lifestyle as well.

Innovations in packaging allowed for the marketing of individual products, so that consumers bought not just a loaf of bread, but a loaf of Wonder Bread. This development ushered in what historian Richard Tedlow refers to as the "golden age" of brand-name products.[18] Brands were not only a manufacturer's best friend, but—important for the present discussion—they helped "democratize American life." Standardized products can be bought and sold to anyone regardless of social standing. Americans joined "consumption communities" in which the common thread was the consumption of the same brands.[19]

Together with advertising, buying on credit is a linchpin of consumerism; indeed, it may be the marketing innovation most responsible for today's consumer culture (and its attendant problems). Although Americans had long been buying groceries at the corner store "on account," the use of interest-accruing credit for consumer purchases was new. The "buy now, pay later" ethos, introduced at the turn of the century, greatly increased consumer spending. Over the period 1898–1916, Americans nearly doubled their purchase of automobiles, pianos, and other high-ticket items. Already by 1924, nearly 75 percent of all cars were bought on credit. According to historian Gary Cross, by 1925, 70 percent of furniture, 75 percent of radios, 80 percent of phonographs, 80 percent of appliances, and 90 percent of pianos were purchased on the installment plan.[20]

Clearly, then, by the turn of the twentieth century, Americans were increasingly embracing the shiny-objects ethos—the notion that the good life is dependent on consumer goods. By 1906, merchant John Wanamaker, founder of the Grand Depot "department store" in Philadelphia (one of the first in the United States), announced that a new culture was emerging in the United States. At its heart, he said, was the "quest for pleasure, security, comfort, and material well-being."[21] William Leach, in his book *Land of Desire*, quotes a contemporary of Wanamaker's who had this to say about the new culture: "It speaks to us, only of ourselves, our pleasures, our life. It does not say, 'Pray, obey, sacrifice thyself, respect the king, fear thy master.' It whispers, 'Amuse thyself, take care of yourself.'"[22]

THE FORDIST DEAL

Although many of us acknowledge Henry Ford as a pivotal player in mass production, few of us have considered that he might also rightfully be considered the father of mass consumption. In the so-called Fordist deal, Henry Ford offered his employees the promise of a living wage and an increasing standard of living in exchange for the repetitive and alienating work of the assembly line—in other words, for being a mere cog in the wheel of the manufacturing process. Researchers Yiannis Gabriel and Tim Lang put it this way: "Ford offered his work force the carrot of material enjoyment outside the workplace as compensation for the de-skilling, control, and alienation that he imposed in the workplace."[23] The true cost of that deal would be played out in years to come in factories and workplaces that reached far beyond the automobile industry.

The lure of the then-incredible sum of five dollars a day for eight hours of work was too much to resist for workers laboring for two or three dollars a day for ten or more hours each day. Ford was a business visionary. He recognized that paying a decent wage turned his employees into potential customers as well. Ford was heard to say, "If you cut wages, you just cut the number of your customers."[24] The shorter workday he imposed allowed workers the time to spend their money and fashion a lifestyle focused on consumption. No longer bound to the land, factory-working Americans were required to live on their wages or salaries, giving money a new preeminence in their lives. Money would now become the base measure of all other values. Under the Fordist deal, a person's identity and meaning were intimately associated with his work and with the standard of living made possible by his wages.

In 1908, the Model T (often referred to as the Tin Lizzie) sold for $950—a price that made it accessible to the burgeoning middle class of the day. Five years later, Ford introduced his version of the assembly line. With the help of that innovation, by 1924 an improved, enclosed Model T could be purchased for only $290! This less expensive model made it possible for nearly any person making a decent salary, even Ford's own workers, to own one—truly a car for the masses. In that vehicle's nineteen-year production run, Ford sold 15.5 million Tin Lizzies. As late as 1910 the United States manufactured only 180,000 autos. By 1924, 4 million cars were produced. By 1927, 85 percent of all of the world's automobiles were made in the United States.[25]

The American consumer society was created by an influx of dramatically new products that became available to the masses in the 1900s. As remains the case today, the automobile—the standout of those products—captured the aspirations of the working classes with its promise of freedom (possibly from the drudgery of assembly-line work!) and a sense that one had found his place within the social milieu. The automobile was, and remains to this day, the most prized commodity of the consumer culture—a symbol of the new American Dream.

THE AMERICAN WAY

The 1920s were a time of great upheaval. Nowhere was this more evident than in the priorities of the American people. Writers of the day, including Sinclair Lewis (*Babbitt*) and F. Scott Fitzgerald (*The Great Gatsby*), depicted Americans as being pulled in two different directions. America was searching for its moral compass and apparently discovered it in the frantic pursuit of material possessions and an ever-increasing standard of living. Unlike the walker depicted in Robert Frost's poem "The Road Not Taken," few Americans hesitated in choosing the material path, and even fewer stopped to consider where their chosen road might lead them.

In 1928 Frenchman André Siegfried wrote, "A new society has come to life in America." Siegfried had visited America four times from 1901 to 1925 and had increasingly found it drifting to an ethos linked to industrialization and mass consumption. "It is obvious," he concluded, "that Americans have come to consider their standard of living as a somewhat sacred acquisition, which they will defend at any price. This means that they would be ready to make an intellectual or even moral concession in order to maintain that standard."[26]

Americans were ready for a party in the 1920s. World War I had ended, and an entire generation of GIs was returning with money to celebrate.

Furthermore, the war machine was quickly being reloaded for the peacetime production of a dizzying array of consumer goods, apparently buying into Karl Marx's assertion, "Without production, no consumption; but also, without consumption, no production."[27] Fortunately for consumers, mass production allowed a wide range of products to be manufactured at prices now affordable by all. One of the most important was the radio—the first mass broadcasting media. Radios were both affordable and entertaining and, with the ads they disseminated, helped develop the national culture of consumption necessary to sell all of the products being produced.

Ten years after radios first became widely available in stores in 1922, 55 percent of American households owned one. Many Americans were paying for theirs on an installment plan. The Radio Corporation of America (RCA) was formed in 1920, and the first national system of radio stations followed in 1926 (NBC), with CBS following the next year. What started out as only a few independent radio stations in 1920 had ballooned by 1930 to over 600 stations regularly broadcasting programming and advertising.[28]

Advertisers took their job as change agents very seriously. They were the self-proclaimed apostles of modernity, civilizers, and, most important, "Americanizers." The industry's overarching goal was to create what we might call a "democracy of desire," because without a broad-based desire to be satisfied, mass consumption could not be sustained. Billions of dollars were spent to create not only the necessary state of desire but also a vision of the new American way of life.

As noted earlier, of all the products that flooded the marketplace, none played a more important role than the automobile. Although Henry Ford and his Model T ruled the auto market in its formative stages, selling 55 percent of all cars in 1921, it quickly lost ground to General Motors, whose primary innovation was the annual style change. Unlike the company that produced the dull and stodgy Model T (any color you want, as long as it's black), General Motors allowed its customers to choose from a wide range of colors and models. These new choices, and various style changes that General Motors introduced annually, made the automobile a source of status among the buying public. And the public responded overwhelmingly. Americans wanted to have a car of their choosing—one that could say something about who they were or what their socioeconomic status was. Already by 1927, Ford's market share had fallen to 25 percent. By the time old Henry relented on annual style changes, it was too late for the Ford Motor Company to make up for lost ground.[29]

Owning an automobile was so important in those early years that people were willing to sacrifice food, clothing, and even savings to own a car. An automobile in the driveway was considered not a luxury, but a necessity. Many families mortgaged their houses to own a car. Most cars—75 to 90 percent—were bought on an installment plan, and the average working man paid approximately one week's wages each month to have the privilege of owning a car.

The economic prosperity experienced during the 1920s helped to securely entrench material possessions and an ever-increasing standard of living as cornerstones of the American Dream. Although very much in the minority, there were those who criticized and lamented America's preoccupation with consumption. Appropriately enough, one such critic, Samuel Strauss, labeled America's love affair with stuff "consumptionism." Consumptionism, argued Strauss, was a life philosophy dedicated to the production and consumption of more and more things—"more this year than last year, more next year than this." In Strauss's thinking, this philosophy emphasized the "standard of living" above all other values. He saw consumptionism as "something new [that] has come to confront American Democracy." Consumptionism had produced, in Strauss's words, "astonishing" changes in the United States. First, Americans no longer reviled "rich men," the Robber Barons of the turn of the century. A newfound respect had replaced the open hostility and mistrust of the captains of industry of only a few short years earlier. Second, Americans were realigning their priorities away from the hard work and sacrifice of the Protestant ethic and toward a pleasure ethos focused on "luxury and security and comfort" as the foundational elements of the "good life."[30]

A DEFINING MOMENT

It wasn't until 1931, in his book *The Epic of America*, that historian James Truslow Adams added the phrase "American Dream" to the world's lexicon.[31] Adams was born in 1878 into a family of limited financial means. He earned an undergraduate degree from Brooklyn Polytechnic Institute in 1898 and a master's degree in philosophy from Yale University in 1900. Following in his father's footsteps, Adams went to work on Wall Street, where he was able to make enough money to pursue his true love—writing. His histories of Long Island garnered him some note, and he was asked by a presidential adviser, Colonel Edward M. House, to collect data for the upcoming Paris Peace Conference following World War I.

After the war, Adams wrote his "New England Trilogy," the first volume of which—*The Founding of New England*—garnered him a Pulitzer Prize in 1922. He gained note as a scholar with his 1927 book *Provincial Society*, which appeared in the prestigious *History of American Life*—a series edited by top academic historians. Clearly, he was a respected scholar by the time he wrote *The Epic of America* several years later.

The intended audience for *The Epic of America* was a general readership; the book was not a scholarly tome. Much to his disappointment, his publishers at Little, Brown and Company did not feel that a book entitled *American Dream*—his own proposed title—would sell. This was clear evidence that the phrase was not yet part of the common vernacular. Well, if the phrase couldn't be the title for his book on the history of America, it would nonetheless play an important role throughout the book: Adams referred to the American Dream thirty times in its pages.[32]

It didn't take Adams long to work that crucial phrase into his book. In fact, it was in the book's preface that Adams defined what he believed to be the most important historical theme, which he called the

> American Dream of a better, richer, and happier life for all of our citizens of every rank which is the greatest contribution we have as yet made to the thought and welfare of the world. That dream

or hope has been present from the start. Ever since we became an independent nation, each generation has seen an uprising of the ordinary Americans to save that dream from the forces which appeared to be overwhelming and dispelling it. Possibly the greatest of these struggles lies just ahead of us at the present time.[33]

It's ironic that this phrase defining a dream of endless opportunity would be coined during the early years of the Depression, when so many saw opportunity slipping away. Despite this dubious timing, the phrase "American Dream" rapidly entered the lexicon both in the United States and abroad. It was Adams, then, who put to paper what many over the years have referred to as our "national motto."

Jim Cullen, author of *The American Dream: A Short History of an Idea That Shaped a Nation*, observes that "beyond an abstract belief in possibility, there is no one American Dream. Instead, there are many American Dreams."[34] Adams's definition of the American Dream in the epilogue to *The Epic of America* describes it as "that dream of a land in which life should be better, richer, and fuller for every man, with opportunity for each according to his ability or achievement."[35] In those words we hear what I consider a more expansive and traditional understanding of the American Dream.

The American Dream, Adams explained, is "not a dream of motor cars and high wages merely, but a dream of a social order in which each man and each woman shall be able to attain to the fullest stature of which they are innately capable."[36] Already by the early 1930s, when he wrote his *Epic*, Adams was concerned that the economic aspects of the American Dream were taking center stage. He worried that Americans were more preoccupied with "how size and statistics of material development came to be more important in our eyes than quality and spiritual values."[37] It seemed to him that Americans had forgotten how to live, in their struggle to "make a living."

Adams's own definition left room for a pecuniary interpretation of the American Dream, however. After all, what does "better" or "richer" mean? Adams minced no words as to how contemporary business leaders would define such terms. In his view, the higher wages being offered to workers were not intended to create more well-rounded people, by increasing their leisure and the chances to make good use of it, but to provide Americans with the time and money to expand their powers as consumers. Adams lamented over what he saw as excessive advertising and other efforts to cajole Americans to spend their wages on consumer goods. Even in the very heart of the Depression, advertising agencies insisted that the country's financial woes were mainly a problem of spending (or lack thereof). As much as things change, it appears they stay the same: both the Bush and Obama administrations encouraged Americans to spend their way out of trouble.

In one of the earliest mentions of the treadmill of consumption, Adams warned of the work-and-spend cycle of consumerism. The American Dream, warned Adams, "can never be wrought into a reality . . . by keeping up with the Joneses. . . . We cannot become a great democracy by giving ourselves up as individuals to selfishness, physical comfort, and cheap amusements."[38] It appears that even the better part of a century ago, the American people had spoken: "better and richer and fuller," in regard to the American Dream, were to be defined in terms of money and material possessions.

FREEDOM FROM WANT

Franklin Delano Roosevelt was elected our thirty-second president, defeating Herbert Hoover, on November 8, 1932. In July of 1932, before his election victory, he had already pledged "a New Deal for the American people," when he accepted his party's nomination for president. By the time FDR took office on March 4, 1933, the nation was in the midst of the Great Depression. Millions of Americans were out of work and hungry, and the nation was reeling. Given the times, FDR's primary focus was to help secure the "freedom from want" for all Americans.

Many of the measures passed during the FDR presidency, including those establishing Social Security, federal unemployment insurance, and a national minimum wage, had a profound impact on how Americans viewed the American Dream. When FDR began to implement such sweeping changes under his New Deal, the American Dream was, as author David Kamp puts it, "maturing into a shared dream."[39] The individualistic, Wild West version of the American Dream was slowly fading away, the end of that era having been heralded by the 1890 Census Bureau proclamation that no real frontier remained. Under FDR's leadership, the American Dream of a "better and richer and fuller" life put to paper by James Truslow Adams in 1931 was, in Kamp's words, "no longer just what America promised its hard working citizens individually; it was an ideal toward which these citizens were duty-bound to strive together."[40] With the passing of the Social Security Act, requiring workers and employers to contribute to federally administered trust funds for the benefit of those lucky enough to make it to a ripe old (and increasingly older) age, Americans were guaranteed—in theory, at least—some measure of economic security in their golden years. This, argues Kamp, may have been "the first time that a specific material component was ascribed to the American Dream."[41]

Although the sweeping changes brought about by the various measures enacted as part of FDR's New Deal helped shape a shared vision of the American Dream, the dream remained a set of loosely articulated but deeply held personal values rather than a clearly articulated list of social or economic goals. FDR would help to transform those personal goals into a national agenda as he worked through his twelve-year presidency. In an effort to steel the American people for World War II, he delivered possibly his greatest speech ever—remarks on what he called "the four freedoms"—tacked onto the end of his State of the Union address in 1941. In those concluding remarks, Roosevelt enunciated four fundamental freedoms that Americans and people around the globe should be entitled to enjoy: the freedom of speech and expression, the freedom of religion, the freedom from want, and the freedom from fear. FDR presented these freedoms, not as guiding principles, but as values of the American people.[42] Of particular interest to the present discussion, the freedom from want went "beyond the traditional American constitutional values protected by the first Amendment of the Constitution . . . endorsing the right to economic security."

Here are the freedoms as Roosevelt outlined them:

> In the future days, which we seek to make secure, we look forward to a world founded upon four essential human freedoms.
>
> The first is freedom of speech and expression—everywhere in the world.
>
> The second is freedom of every person to worship God in his own way—everywhere in the world.
>
> The third is freedom from want—which, translated into universal terms, means economic understandings which will secure to every nation a healthy peacetime life for its inhabitants—everywhere in the world.
>
> The fourth is freedom from fear—which, translated into world terms, means a world-wide reduction of armaments to such a point and in such a thorough fashion that no nation will be in a position to commit an act of physical aggression against any neighbor anywhere in the world.
>
> That is no vision of a distant millennium. It is a definite basis for a kind of world attainable in our own time and generation. That kind of world is the very antithesis of the so-called new order of tyranny which the dictators seek to create with the crash of a bomb.

In the first three freedoms shared by Roosevelt, we see the primary core values of the American Dream: the freedom of speech and expression, the freedom to worship God in our own way, and—on a more worldly plain—the freedom from want; these are all essential ingredients of a "better and richer and fuller" life. Think of these components of the American Dream as the supports of a three-legged stool. Each must be carefully guarded. If any leg is shortened or lost, the American Dream is compromised.[43] FDR also addressed, in that speech, the importance of economic stability, saying that the American people should be able to expect "the enjoyment of the fruits of scientific progress in a wider and constantly rising standard of living"[44] and that increasing numbers of citizens should be covered by old-age pensions and unemployment insurance. Say what you might about FDR's politics, he did more than any other president (before or since) to help clarify the American Dream, and he rallied the American people in its pursuit.

FDR's four freedoms speech was given support by an unlikely ally—painter Norman Rockwell. Moved by FDR's speech, Rockwell painted a set of four freedoms paintings that were published in the *Saturday Evening Post* on February 20, February 27, March 6, and March 13, 1943. Distraught by the fact that he was too old to serve in the military, Rockwell sought to serve his country by illustrating each of the four freedoms.

Rockwell created his now iconic paintings over a six-month period in 1942. He was so obsessed with the project that he lost fifteen pounds and had many sleepless nights while creating possibly the greatest masterpieces of his illustrious career. Ironically, Rockwell turned to his friends at the *Saturday Evening Post* and Curtis Publishing only when no United States wartime government agency—not even the Office of War Information (OWI)—would sponsor the paintings.

The public response to the paintings in the *Post* was phenomenal. Realizing that it had missed a golden opportunity to support the war effort, the U.S. Department of the Treasury (with the help of Curtis Publishing) organized a nationwide tour of the four freedoms paintings, called "The Four Freedoms War Bond Show." The tour took the original four paintings to sixteen cities across the United States, where more than a million people viewed them. The paintings struck a chord with the American people—putting to canvas feelings and values at the core of the American Dream.[45]

The freedom from want painting was the most popular of the four, possibly given the difficult times. The painting depicts a happy extended family sitting down for Thanksgiving dinner. Though the plattered turkey is large, no excess or ostentation can be found in the painting. Over-the-top holiday decorations are noticeably absent, as is the holiday food orgy typically found in America today: only a bowl of fruit and celery adorns the simply appointed table, and everyone is drinking water. The home also hints that it is modest and minimally decorated—unadorned white curtains frame the window.[46]

The Office of War Information, after initially expressing a lack of interest in the paintings, finally grasped their power to stir the American people and printed 2.5 million copies for distribution across America. The four freedoms tour sold over $130 million of war bonds, which provided much-needed funds and helped shorten World War II—further support for the old adage that a picture is worth a thousand words (or more!).[47]

At that point in our nation's history, as today, the American Dream was still a work in progress. Chapter 3 continues the story of the evolving American Dream, examining how such events as the rise of the consumer ethos of the 1950s, the turbulence of the 1960s, Jimmy Carter and the 1970s, Ronald Reagan and the 1980s, the Internet bubble, and the mortgage crisis of the twenty-first century all helped shape, for better or worse, the uniquely American way of life.

THE AMERICAN DREAM ON STEROIDS

Too many of us look upon Americans as dollar chasers.
This is a cruel libel, even if it is reiterated thoughtlessly
by the Americans themselves.

—ALBERT EINSTEIN

On July 24, 1959, Vice President Richard Nixon had a series of impromptu exchanges with Soviet Premier Nikita Khrushchev. The occasion for the somewhat heated debate was the opening of the American National Exhibition in Sokolniki Park in Moscow. At that exhibition, an entire house, cut in half so the insides could be revealed (and more likely ogled), was on display to showcase American technology, ingenuity, and, most importantly, standard of living. Nixon described the exhibition proudly: "Let us start with some of the things in this exhibit," he said. "You will see a house, a car, a television set—each the newest and most modern of its type we produce. But can only the rich in the United States afford such things? If this were the case, we would have to include in our definition of rich the millions of American wage earners."

The American system worked, Nixon affirmed, stating that "44 million families in America own 56 million cars, 50 million television sets,

[and] 143 million radio sets, and 31 million of those families own their own homes." He also boasted about the vast array of other products available to the American people, including dishwashers, lawn mowers, cars of many makes, colors, and styles, TV dinners, makeup, lipstick, spiked heels, hi-fi sound systems, and even Pepsi-Cola. His emphasis on such household appliances as the dishwasher led to the exchange being called the "Kitchen Debate." Like his "fellow Americans," Nixon saw the material things as representing the American Dream. As historians Richard Fox and T. J. Jackson Lears note, Richard Nixon's "Kitchen Debate" with Khrushchev "was a pure expression of the times. The American way of life equated with the American standard of living."[1]

THE OPIATE OF THE MIDDLE CLASSES

Freedom and opportunity have always undergirded the American Dream— but the freedom and opportunity to do what? Postwar Americans had a simple answer to that question: the freedom and opportunity to buy what they wanted, not what they needed—in other words, to become the world's überconsumers. Pent-up demand fueled the consumer frenzy that followed World War II. That war had lifted the United States out of the Depression, but it had brought with it its own consumer hardships, in the form of rationing. Despite shortages of particular foods and products, money seemed to be flowing: the federal budget rose from $9 billion in 1939 to $100 billion by 1945, and much of this money fell into the hands of the American people. Stateside workers often worked overtime in the factories, and their purses and wallets brimmed with cash. This, for many Americans, was the first time in nearly a decade that they were confronted with the question of what to do with their extra money.[2]

The problem was that there were not a lot of places to spend their largesse. The military needs of the country came first, which dried up the supply of many consumer goodies including, famously, chocolate and nylon stockings, but also radios and cars and even the gas needed to drive a car. In addition, there was a housing shortage as construction workers were diverted to wartime projects. All in all, the American people were living an austere lifestyle—and not of their choosing. By the end of 1944, Americans had amassed savings of $140 billion.[3] Money was burning a proverbial hole in their pockets, and the American people looked forward to better days on the horizon. Those days would come with the end of the war at the Armistice of August 14, 1945. America hardly blinked an eye before converting war technologies and production to consumer mass production. The full flowering of consumerism following World War II represents one of the greatest changes in human experience the world has ever seen. The vast majority of Americans began to define life somewhat differently after the war—that is, with a decidedly consumer bent.

The promise of a consumer cornucopia in postwar America was not something entirely new; in essence it was a resurrection of the consumer ethos of the 1920s. The confluence of major retail stores, the mass production of consumer goods, and the emerging advertising industry in the 1920s planted the seeds for the full-blown consumerism to come. These developments had subtly changed personal behavior and values—particularly the value of thrift, which is critical to how we spend money. The subsequent Great Depression and World War II certainly slowed the emergence of the consumer culture, but only temporarily.

"No man who owns his own house and lot can be a communist," quipped the point man for the postwar housing boom, William Levitt.[4] "Levittown," now part of the American vocabulary, is a label used to represent the clusters of tract-style homes that fueled the boom. By today's standards, the houses Levitt built were cramped. The first ones, built in 1947, were approximately 800 square feet and were characterized by their cookie-cutter construction. With row upon row of the same house, it was difficult to distinguish one from the next.

Levitt had been a Seabee—a member of the Navy's engineering corps—in the Pacific during the war. As a member of one of the Navy's construction battalions, where he was responsible for building airstrips as quickly and economically as possible, he learned the art of mass production. That training, on top of his experience in his father's construction business prior to the war, imbued Levitt with the skills needed to quickly build affordable housing for the throngs of returning veterans. Signed into law by FDR in 1944, the GI Bill allowed returning veterans easy access to homeownership.[5] It didn't hurt that there was an existing shortage of housing. Between the years of 1947 and 1957, veterans snapped up nearly 50 percent of all home loans. And the homes were priced to move: $7,900 for two bedrooms and $9,500 for the "spacious" three-bedroom model. The homes had no basements and few had garages, but many did come with what then would be considered luxuries including fireplaces, household appliances, and—for a lucky few—built-in televisions. (You were pretty much the first on your block if you had a television in the late 1940s.) In 1940, only 603,000 new homes were built, but by 1950 a staggering 1.95 million homes were constructed. The construction rate continued at approximately 1.22 million new homes built each year throughout the 1950s.[6]

William Levitt's ability to mass-produce affordable housing, his expansion beyond the borders of the state of New York, and the work of a growing flock of construction competitors secured homeownership as an additional tenet of the American Dream, which continued to transform itself from the "richer" and "fuller" aspirations articulated by Adams to the more specific act of homeownership. But by no means were the American people ready to stop there. New goods, in the collective mind of the American people, represented progress. To consume was to be free.

The American Dream of the postwar years can be neatly summed up in the phrase "standard package." The ranch, Cape Cod, or split-level home in the suburbs was only the beginning. The "embourgeoisement" of the American people, whereby the working classes became part of the bourgeoisie, would not be complete without an entire complement of necessary goods.[7] Sociologist David Riesman refers to this trend as the "career" of consumption. Americans of that era, observes Riesman, had a distinct image of the desired American lifestyle—a lifestyle characterized by "a set of goods and services including such household items as furniture, radios, television, refrigerator, and standard brands in food and clothing."[8] This was the standard package that allowed one to enjoy all that is American—to partake in the American Dream.

The purchase of a home in the suburbs and a car in the driveway, the "foremost" symbol of the age, led the consumer charge. In 1954, Americans bought nearly 5 million new cars—used simply wouldn't do. In 1955, they bought nearly 7 million cars, an increase of 35 percent from the year before.[9] Despite the burgeoning sales of homes, automobiles, blenders, refrigerators, furniture, radios, phonographs (remember those?), and the myriad of gewgaws available to the American public, none played a more important role in defining the American Dream than the television.

Television, according to historian Gary Cross, "was easily the most important domestic consumer product in the 1950s."[10] Although the technology for the television was already in place much earlier, the Great Depression and World War II placed the product's introduction onto the social back burner. Once introduced, however, its impact on the fabric of America was profound. In 1950, only 9 percent of American households had a television set; phenomenally, by 1960, nearly 90 percent of all households owned (or were paying for) a set.[11] I enjoy listening to my mother-in-law talk about how friends and neighbors would flock to her home to watch comedies like *The Life of Riley* and variety shows like *The Ed Sullivan Show*. Friends would often sneak over during the day for a cup of coffee and a peek at popular soap operas.

The advent of television struck a crippling blow to both moviegoing and radio listenership. Between 1946 and 1953, the audiences for both movies and radio dropped by nearly half. By 1960, the average American household watched five hours of TV every day. America's love affair with television had begun.[12] And all that viewing permitted advertisers unprecedented access to the American public. Cross, in his book *An All-Consuming Century*, may have said it best: "Television became a nearly perfect expression of suburban life . . . while enticing viewers through commercials to the miracle miles of fast food chains and shopping malls."[13]

Culturally, television had an enormous impact. It helped homogenize the American way of life. David Kamp, in a *Vanity Fair* article entitled "Rethinking the American Dream," observed that "nothing reinforced the seductive pull of the new, suburbanized American Dream more than the burgeoning medium of television."[14] Programs like *The Donna Reed Show*, *The Adventures of Ozzie and Harriet*, *Father Knows Best*, and *Leave It to Beaver* provided a squeaky clean (and highly idealized) look at middle-class suburban living in America. Still, Kamp notes, "the American Dream was far from degenerating into the consumerist nightmare it would later become (or, more precisely, become mistaken for)."[15] If you need to be reminded of the relative modesty of early TV programming, simply find a rerun of any of the above shows currently running on television. Nowhere will you find the hot tubs, "great rooms," fancy kitchens, and triple-car garages that are standard in many homes today.

By the 1950s everything was in place—the stage was set for the maturing consumer culture: governmental policies such as Social Security, low-income housing loans, and credit legislation; high wages through the growth of unions that enhanced consumer capacity; mass media (including television) that stoked the ideological flames of consumerism; expansion of lending institutions; the Baby Boom generation; and, importantly, the changing attitudes toward thrift and affluence. Americans now saved, not to accumulate wealth, but to spend—and this spending was often accompanied by consumer debt.[16]

The "standard package," notes historian David Steigerwald, would not have been possible if "Americans [had] maintained their commitment to thrift."[17] Many Americans, observes professor Jan Logemann, "came to regard credit as a means of ensuring democratic access to the American Dream and to an expanding middle class."[18] Typically, it was young suburbanites who were using installment loans and other forms of credit to achieve middle-class standing. Credit was seen as a means of "social climbing." Keeping up with the Joneses came at a price, however. Eminent psychologist Ernest Dichter concluded that "America is experiencing a revolution in self-indulgence. . . . We've learned that one rarely makes one's ultimate goal—so why not enjoy life now?"[19] A somewhat unsettling observation, but one that illuminates the seismic shift that took place in how Americans viewed work and thrift, and thus the American Dream.

The use of credit became nearly universal when the last vestiges of control were lifted with the conclusion of the Korean War. The removal of the so-called Regulation W, which had placed restrictions on consumer credit during World War II, led to much more lenient credit terms. Prior to this—horror of horrors—an auto purchase required one-third minimum down payment, and the remaining credit could not extend past fifteen months. Even major appliances required 20 percent down and had to be paid off in no more than twelve months. During World War II, store charge accounts typically had to be paid off in six months. As Steigerwald puts it, "1954 can be used as something of a date for contemporary society's coming-out party, the point at which the system was fully, and finally, unleashed."[20]

The home was the nexus that connected the cars, televisions, furniture, and myriad appliances. It was the nerve center of all spending and consumer debt, because mortgage debt helped make people more comfortable with debt in general. The 1950s saw all types of consumer debt increase by a startling 200 percent. In the period spanning 1929 to 1955, Americans spent approximately six times more on personal loans while installment borrowing increased from $3.1 billion to $29 billion. By the end of the 1950s, a whopping two-thirds of American households were saddled by some type of debt.[21] As noted earlier, the availability of installment plans was critical to the emerging consumer culture.

Still, because of increasing wages during the 1950s and '60s, Americans were able to save at about a 7 percent rate. In fact, Americans banked approximately $14 billion in 1952—the second-highest total amount of savings in any year to that point. In 1955, a real boom year for consumer borrowing, Americans still managed to pay down debt and save almost $12 billion.[22] Despite ongoing saving, the 1950s and '60s saw Americans become first comfortable with debt and then dependent upon (addicted to?) it—even after our borrowing outstripped our ability to repay.

"HELL NO, WE WON'T GO"

In contrast to the relatively tranquil 1950s, with the beatniks' cries of "Slow death by conformity" leveled at the burgeoning middle class, the 1960s were a decade of great upheaval. The war in Vietnam, civil rights protests, evolving women's rights, and sex, drugs, and rock 'n' roll all became battlefronts during the turbulent 1960s. Moreover, a real questioning of an American Dream based on consumerism had begun. Many of the young people involved in the counterculture movement of the 1960s wanted out of what they perceived as the "rat race" so wholeheartedly embraced by their parents. Timothy Leary—Harvard professor, proponent of LSD experimentation, and leading light of the counterculture movement—rallied the troops to "turn on, tune in, and drop out." The faithful were not to work for change within the current system, but to reject the system, drop out of the "corporate network of consumption," and reject the "culture of malls and surburban status seeking."[23]

Such sentiments were not the sole dominion of the beatniks (and later hippies), however. This revolt against the growing materialism of the suburban consumer culture had its vocal mainstream proponents. Vance Packard's best-selling books *The Hidden Persuaders* (1957), *The Status Seekers* (1959), and *The Waste Makers* (1960) argued that Middle America was taking its cues from advertisers, television, and the movies. The new-found mass affluence of the 1950s and '60s had not created a classless society. Rather, as Gary Cross puts it, it had "produced a mass of insecure individuals each trying to define and display themselves through their goods."[24] Packard pleaded with the American people to eschew the status game and listen to their inner voice. Unlike sociologist Thorstein Veblen, who at the turn of the nineteenth century attacked what he called the "conspicuous consumption" of the moneyed elite, Packard directed his comments at the lower rungs of the socioeconomic ladder—the growing middle and lower-middle classes.

Writing in the 1960s, influential German philosopher Herbert Marcuse saw an American middle class that was not exploited by their employers, but had become "happy slaves" who believed that the system would provide for their needs. Marcuse's biggest concern was that Americans were looking for gratification only within the consumer culture, "which, in satisfying [the consumer's] needs, perpetuates his servitude."[25] Author and professor Juliet Schor refers to this perpetuated state as the "work-spend cycle." People are working to buy the things they want, requiring more work, and thus causing the cycle to repeat itself ad infinitum. We work to buy the things we're told we need—you fill in the many blanks here—and are enslaved to meaningless jobs in their endless pursuit.[26]

The counterculture considered anathema the "happy slaves" trade-off of boring and meaningless work for the trappings of the consumer culture. The self-proclaimed leader of the counterculture, Jerry Rubin, offered this bon mot to the youth of America: "Our message: don't grow up. Growing up means giving up your dreams."[27] Such sentiments fell on fertile ground within the youth movement but were widely denounced by the larger culture. Looking back, we can see that consumerism was a fait accompli; eventually even the most ardent countercultural proponents would be swept up by the strong consumer current. Case in point: after years of "searching," Jerry Rubin had this to say about money, "It's O.K. to enjoy the rewards of life that money brings."[28] Regardless of the ultimate failure of the counterculture movement, it was a serious denunciation of the consumerist version of the American Dream that had been evolving in earnest since the middle of the nineteenth century.

Eventually, however, the Baby Boomers of the 1960s would have to grow up, possibly graduate, find jobs, start families, and buy into (literally) the values of the prevailing culture. The old adage that you can't fight city hall comes to mind. Marcuse may have said it more eloquently when he mused, "Can the outcome, for the near future, be in doubt? The people, the majority of people in the affluent society, are on the side of that which is—not that which can and ought to be."[29]

The 1960s protest cries of "Hell no, we won't go!" clearly had nothing to do with shopping. The Vietnam War took the brunt of the protesters' anger. Despite considerable antiwar feeling and the social turmoil of the 1960s, the consumer machine raged on. Americans were enjoying the material trappings of a growing economy and could not envision a world without an ever-increasing pile of material possessions.

Even the counterculture itself couldn't make a clean break from the culture of consumption. People seemed to need particular products and apparel to communicate with others and to feel part of a larger group—hence the tie-dyed look, go-go boots, and jeans. But styles have a way of crossing dividers: just as the hip-hop styles of today have spread beyond that particular subculture, Middle America was soon wearing the blue

jeans of the protesting youth movement. Eventually the entire counterculture would be absorbed by the mainstream.

The decade of the 1960s saw personal consumption increase by 82
percent. As Gary Cross notes, "By the early 1960s, consumer surveys
showed that Americans wanted a more varied basket than the standard
package of car, house, and furnishings."[30] Many purchases led to further spending. Once autos became more affordable, for example, there
was money left in the budget for such things as vacation homes or pleasure boats. Cars had lost some of their panache as status symbols, but
other acquisitions could give one a leg up on the neighbors. By 1960
there were already 2 million families with second homes, and interest in
boating had mushroomed. Forty million Americans took to the lakes—
doubling the 20 million that considered boating as a chosen pastime in
1950. Overseas air travel and car trips likewise increased exponentially.
The percentage of families who owned two or more cars nearly doubled
to 29 percent by 1970. According to Cross, the ratio of Americans to
cars dropped from 3.74 Americans per car in 1950 to 2.9 in 1960 and
1.86 in 1980. And forget about one TV per household. Any upstanding
family might have a portable in the kitchen or bedroom along with the
big set in the family room. In the 1960s, spending again drove the economy: consumers were purchasing appliances, color TVs, autos, furniture, and a myriad of leisure-oriented products such as pool tables, hotel
stays, airline tickets, and games and gadgets for grownups and kids
alike.[31]

It's probably not purely chance that the 1960s saw the emergence of
the enclosed shopping mall. Early versions of today's sprawling complexes, large shopping malls in the mid-1960s offered up to a million
square feet of retail space along with such non-retail tenants as hotels,
movie and office complexes, and a myriad of other attractions. In addition
to the Mall of America, arguably the most visited mall in the world and
the largest in the United States, Minnesota boasts the first postwar enclosed mall in America: Southdale.

A particularly telling event of the 1960s was the repealing of the "blue laws" that in many communities had banned shopping on Sundays. Discount stores, groceries, and drugstores had already become dependent on Sunday business, cutting into the sales of department and specialty stores. Reluctantly, in 1969 Sears cried uncle and opened for Sunday business. In what seems quaint by today's standards, the company ran newspaper ads expressing regret about opening on Sundays, appealing to their competition to remain closed on Sundays and "give our employees their Sabbath." Market forces, however, were too strong, and soon Sunday had become prime shopping time. Such developments as enclosed shopping malls and evening and Sunday hours made it a little easier for Middle Americans to spend their hard-earned money.[32] According to Cross, by the 1970s, Americans spent four times as many hours shopping as their European counterparts.[33]

MAINTAINING THIS TORRID SPENDING SPREE REQUIRED THAT HUSBANDS take second jobs, more women enter the workforce, and Americans take on a whole lot of debt. In addition, credit use reached new heights in the 1960s. As noted earlier, long before the onslaught of credit cards in the late 1960s, credit played an important role in the typical consumer's finances. A majority of Americans by the 1960s had favorable attitudes toward, and themselves used, installment credit. Stated bluntly, by the 1970s, Logemann notes, "credit had come to be seen as a democratic right for American consumers."[34] Industry proponents of credit argued that the lack of access to credit was a form of discrimination, which excluded millions of Americans from the products "which they need and to which they have a right."[35] Although only 2 million credit cards were in circulation by 1968, soon access by the masses to credit cards would open the floodgates of consumer debt. According to historian Louis Hyman in a recent essay entitled *Debtor Nation: How Consumer Credit Built Postwar America*, "Americans learned to borrow in the midst of prosperity, but only learned the consequences of that borrowing after the prosperity ended"—that is, in the 1970s.[36]

SON OF A SHOE SALESMAN

To truly understand the optimism that would come to surround the 1980s, Ronald Reagan, and his namesake Reaganomics, we must take a quick look back at the 1970s. The 1973 Arab oil embargo and the energy crises of 1973 and 1979 led to high levels of inflation throughout the 1970s and into the early 1980s. Skyrocketing oil prices made it necessary for numerous American businesses to raise their prices. Most Baby Boomers can conjure up vivid memories of long lines at the gas pump, gas rationing, and "closed" signs in the front of gas stations that had run out of gas or hadn't gotten their regular allocation of the stuff that Americans were dependent upon.

Key Economic Indicators: Early 1980s vs. 2009

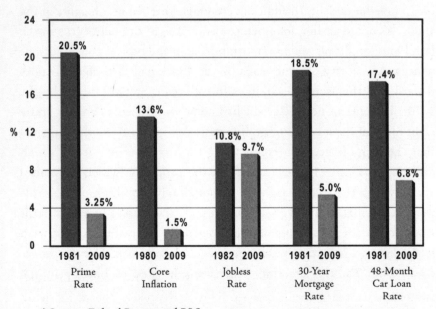

* Sources: Federal Reserve and BLS.

"Stagflation" became part of our lexicon in the 1970s. It's the unenviable combination of high inflation, a stagnant economy, and high unemployment. Interest rates went through the roof. The current (and supposedly fading) recession has often been compared to the Great Depression, earning it the sobriquet the Great Recession. This is all fine and good, but it overlooks the terrible economic conditions of the 1970s and early 1980s. As far as pain and suffering go, the recession of 2008 is not in the same league as that earlier economic downturn. As the graph above attests, the economy of the early 1980s was much worse than our current economic state. Professor Mark J. Perry of the University of Michigan compiled these statistics to stanch the complaining regarding our present economic conditions.[37]

The numbers are mind-boggling. I was a new college graduate in 1982 and didn't really give much thought to the 12 percent savings rate available to me as an employee of Norwest Financial in Golden Valley, Minnesota—the consumer loan division of Norwest Banks. I gave nary a thought to the poor customers charged 20-plus percent for a loan. The prime rate in 1981 stood at 20.5 percent, compared to 3.25 percent in 2009. Inflation stood at 13.6 percent (can that be right?) in 1980—1.5 percent in 2009. The average inflation rate ran 2.5 percent from 1900 to 1970. In the 1970s, inflation averaged 6 percent, with a high of 13.3 percent in 1979. Lots of people are out of work today, but still fewer than the 10.8 percent jobless rate of 1982. The "Misery Index" (the sum of the unemployment rate and the inflation rate) had reached an all-time high of 21.98 percent by the time Jimmy Carter was running for reelection against Ronald Reagan in the mid-1970s.

To put it simply, the Carter presidency had more than its fair share of roadblocks. The economic problems experienced in the 1970s had created a cynical attitude in the American people, accustomed to optimism in the 1950s and even 1960s. Faith in the government was at a nadir, given the dire economic circumstances, the fallout of the Vietnam War, and the Watergate scandal that helped to bring down Nixon. With these events building on the earlier assassinations of John F. Kennedy, Robert Kennedy, and Martin Luther King Jr., the nation was in a foul mood. As President

Carter's respected pollster Patrick Caddell informed him, "For the first time, we actually got numbers where people no longer believed that the future of America was going to be as good as it was now. And that really shocked me, because it was so at odds with the American character."[38] The American people were losing confidence in the American Dream. Americans, argued Caddell, were suffering from a general "crisis of confidence." The energy crises had erupted, and something had to be done to restore faith in the American Dream. Carter's hope of reelection depended upon it. The president's approval rating swooned to 25 percent— lower than Richard Nixon's during the Watergate debacle. It was go-time for the Carter presidency.

To address this crisis of confidence and restore his flagging support, Carter canceled his planned fifth major speech on energy and convened a meeting at Camp David of dozens of prominent Americans, including members of Congress, governors (Bill Clinton of Arkansas was in attendance), labor leaders, academics, and men of the cloth. The purpose of the meeting was to figure out what needed to be done to restore confidence in the future of America and the office of the president. Sitting on the floor and taking notes, Carter (and his administration) bore the brunt of much of the criticism of that crisis. After much thought and deliberation, President Carter was ready to face the American people and inform them what he had learned from his Camp David conclave and what his plans were for extricating America from the crisis.[39]

In the evening of July 15, 1979, with millions of Americans glued to their television screens, Carter gave what would turn out to be his swan song. Since his inaugural address, Carter had stressed that more is not always better, that even a great nation had limits. But it was on this fateful night, in what Caddell referred to as Carter's "Malaise" speech, that Jimmy Carter seriously overestimated the willingness of the American people to sacrifice—to simply go without. In what sounded as much like a sermon as a speech (he was a Baptist Sunday school teacher, after all), he offered the American people the chance at redemption—an offer that had few takers.

And yet Carter was spot-on when he told the American people,

> In a nation that was proud of hard work, strong families, close-knit communities, and our faith in God, too many of us now tend to worship self-indulgence and consumption. Human identity is no longer defined by what one does, but by what one owns. But we've discovered that owning things and consuming things does not satisfy our longing for meaning. We've learned that piling up material goods cannot fill the emptiness of lives which have no confidence or purpose. . . . This is not a message of happiness or reassurance, but it is the truth and it is a warning.[40]

Carter clearly saw the problem, but he failed to appreciate the American people's dedication to consumerism. As the 1980s would attest, Americans were perfectly content with the mindless pursuit of material possessions; it's only that inflation and falling income made such pursuits difficult. Carter had more bad news to deliver, telling the American people,

> We are at a turning point in our history. There are two paths to choose. One is a path I've warned about tonight, the path that leads to fragmentation and self-interest. Down that road lies a mistaken idea of freedom, the right to grasp for ourselves some advantage over others. That path would be one of constant conflict between narrow interests ending in chaos and immobility. It is a certain route to failure. [The other path is the] path of common purpose and the restoration of American values.[41]

President Carter seemed to have a crystal ball that allowed him to see into the future. Unfortunately, the American people did not appreciate being reminded of their shortcomings. Maybe, just maybe, they thought, the problem lay not with the American people, but with its leadership.

You could hear the final nail being pounded into his reelection coffin when he proposed "a bold conservation program" that would enlist "every average American" in overcoming the energy crisis. Carter asked Americans to "take no unnecessary trips, to use carpools or public transportation whenever you can, to park your car one extra day per week, to obey speed limits, and to set your thermostats to save fuel."[42] Few Americans carpooled in the 1970s, and any conservation efforts stopped when fuel prices dropped in the 1980s. Simplicity was exactly what the nation needed, but Americans wanted nothing to do with it. A little over a year later, President Carter lost his reelection bid by a landslide.

Enter Ronald Reagan, whose optimistic vision for the future was in stark contrast to Carter's pessimism and call for sacrifice. Indeed, optimism was Reagan's dominant character trait. He liked to remind the American people that they had always achieved the seemingly impossible, and he rallied them to renew their pride in the United States, announcing that "America is back and standing tall." In his campaign for the 1980 presidency, Reagan downplayed the importance of the energy shortages of the 1970s and accused his rival of overregulation in dealing with them. His goal was to increase the country's energy output to keep the economy going. America's future, according to Reagan, was one of boundless limits.[43] He subscribed, as had Theodore Roosevelt before him, to the doctrine of American can-do-ism. Promising lower taxes (a tenet of supply-side economics) and less government, Reagan won election easily over his more pragmatic and somber rival.

It was clear from the start of the new presidency that Reagan's vision of the American Dream had a decidedly materialistic bent to it. Reagan believed in the promise of the American Dream, but this was no New Deal version of it. In a speech repeated often on the campaign trail, Reagan stated, "We need true tax reform that will at least make a start toward restoring for our children the American Dream that wealth is denied to no one, that each individual has the right to fly as high as his strength and ability will take him."[44] His belief that we are all free to participate in the American Dream to the extent that our abilities will carry us is reminiscent of James Truslow Adams's original definition of the American Dream. The word "entitlement" (and the concept behind it) was nowhere to be found in Reagan's vision for America. But as for "strength and ability,"

Reagan was an icon. Indeed, as the son of a shoe salesman, he was the living embodiment of the American Dream.

The tone for the Reagan presidency was clearly communicated in his 1981 inaugural celebration, a sparkling product of the shiny-objects ethos. Believing that wealth was no longer to be hidden, he spent approximately $11 million of this country's money on the extravagant event. Who can forget Nancy Reagan's $15,000 ball gown or Reagan himself, dressed to the nines in a tuxedo (in stark contrast to Jimmy Carter with his cardigan sweaters and lowered thermostats)?[45] In Carter's inaugural parade, and in keeping with the austere times, the former peanut farmer from Georgia eschewed the presidential limousine, instead walking hand in hand with his wife, Rosalynn, and daughter, Amy. Not so Ronald Reagan, who seemed, with his wife, to revel in the embellishments of the moment. Hollywood stars were everywhere, and "Old Blue Eyes" (Frank Sinatra, for you youngsters) provided the entertainment. The whole inauguration celebration was a glitzy affair. Further cementing the renewed optimism of the 1980s and the Ronald Reagan presidency was the release of the U.S. embassy hostages in Iran only minutes after the new president's inauguration. Expensive renovations of the White House, orchestrated by Nancy, and the purchase of a $200,000 set of White House china set the tone for the remainder of the 1980s. It appeared that it was, indeed, "morning in America."[46] The future, according to Reagan (and his 1984 television commercials titled "It's Morning in America"), painted a bright picture of days ahead for the American people.

DOT-BOMB

In the 1990s and early 2000s America entered into what David Kamp refers to as the "Juiceball Era of the American Dream—a time of steroidically outsize purchasing and artificially inflated numbers."[47] The dot-com boom of the mid and late 1990s was just what the doctor ordered to satisfy Americans' desire to achieve the American Dream with as little inconvenience as possible. Despite the nearly 150 years that separated the two events, the dot-com boom and the Gold Rush shared much in common.

First and foremost, both transformed the American Dream. Looking back at the Gold Rush, historian Jim Cullen explains, "The notion that transformative riches were literally at your feet, there for the taking, cast a deep and lasting spell on the American imagination."[48] That mid-nineteenth-century prospect of easily available riches left an indelible imprint on the American people. Now fast-forward to the mid-1990s. Once again, the lure of easy money captured the imagination of the American people, as will be discussed below.

Second, both were short-lived. Although the Gold Rush would slowly dissipate over the span of approximately a decade, the Internet boom ("Bubble 1.0," some call it) lasted less than five years. In both cases, the aftereffects would last far longer than the events themselves.

Third, both events caught the attention of the nation and the world. Fortune hunters from around the globe came to America to pan for gold. Sad to say, however, few struck it rich. Similarly, the Internet explosion enticed millions of people in this country and abroad to work late or quit their day jobs in hopes of striking it rich on this new thing called the World Wide Web. Ordinary people started new companies from their garage, often investing their limited life savings in hopes of becoming millionaires. A few of the creators of the Internet did it because it was a challenge, certainly, or to help the world communicate more effectively. Not these relative Johnny-come-latelies, though. The race was on, and there was money—lots of it—to be made. The first widely used browser software, Mosaic, was introduced in 1993. It provided a user-friendly (for that time period) interface and compelling graphics and images which opened a world of possibilities for the earliest denizens of the Internet.

But the Internet frenzy wasn't limited to those interested in staking a digital claim. The rest of us computer-challenged Americans were also caught up in the dot-com boom. Stories of untold millions made (but not necessarily earned) in such ventures, and reflected in soaring stock prices and spectacular IPOs (initial public offerings) for Internet companies, had even the most jaded investors, small and large alike—with the exception of the Oracle of Omaha, Warren Buffett—plunking our money down in hopes of taking a shortcut to the American Dream.

To those of us in the trenches at the time of the Internet boom, it appeared that there was big money to be made, and we wanted a piece of the action. As Sally once said to her brother, Charlie Brown, "All I want is my fair share. All I want is what's coming to me."[49] The instant millionaires we read about inspired us. Consider Mark Cuban, who could be considered the prime exemplar for the potential of the Internet. In 1995, with good friend and fellow Indiana University graduate Todd Wagner, Cuban started a company then called AudioNet. Their shared interest in college basketball and Web-casting was the catalyst behind the company. By 1999 the company had been renamed Broadcast.com and had grown to 330 employees. It generated a few million in revenue but posted a loss of $15 million for that year. In keeping with the Internet hysteria of the time—more money than brains chasing dot-coms—Yahoo! purchased Broadcast.com for $5.9 billion (yes, that's billion) in Yahoo! stock.

Wasting little time, Cuban used a small portion of his Internet fortune—$285 million—to buy the Dallas Mavericks basketball team. He also bought a $40 million Gulfstream V business jet in October of 1999, which was identified by the *Guinness Book of Records* as the "largest single e-commerce transaction." His 24,000-square-foot house in the swanky Preston Hollow area of Dallas was bought for $13 million—chump change for such an Internet icon.[50]

Internet fever was spread by news coverage of spectacular success stories such as this, talk over the water cooler at work, and the chutzpah with which the companies promoted themselves. Nowhere was this chutzpah more evident than at Super Bowl XXXIV in January of 2000. I remember it well because every year I have my advertising students rate Super Bowl ads. In what was the dot-com bubble's Waterloo, seventeen dot-com companies paid an average of $2.1 million dollars for a thirty-second spot! Remember the talking hand puppet for Pets.com, a company that spent millions in Super Bowl ads but failed shortly after the dot-com bubble burst? How about the Monster.com ad for its career/job search website, an ad that challenged viewers to take stock of their careers? Their multiple "when I grow up" spots featured children musing about what they wanted to do when they grew up. One child said, "I want to file all day."

Another, "I want to claw my way up to middle management." Or, changing companies, how about chimpanzees dancing the mambo to "La Cucaracha" to promote whatchamacalit.com? Super Bowl viewers that day received a heavy dose of Internet fever—over $40 million of venture capital and stockholder cash poorly spent. Although the commercials were highly entertaining, it was unanimous among my students that most of them had no place on the Super Bowl and that the end must be near. That collective prediction proved prescient.

I wish I could say the same for the corpus of the investing public. Most of us were burned by the dot-com bust. The only question that remained was how badly. The majority of us were left licking our collective wounds when the bubble burst on March 10, 2000, less than five years after it began. Many say that Netscape's IPO on August 9, 1995, marked the start of the Internet boom. Shares opened at $28 and ended the day at $75—a modest increase compared to many later Internet IPOs. By December of 1995, Netscape was trading at $171 per share. Netscape wasn't the first Internet-based company on the market, but it was the first to garner broad media attention.[51]

After hitting an intraday high of 5,132.52, the NASDAQ exchange laden with technology stocks closed at 5,048.62 on March 10, 2000. This was more than twice its value only the year before. Massive, multi-billion-dollar sell orders of bellwether high-tech stocks like Cisco, Dell, and others on the Monday following the March 11–12 weekend knocked a full 4 percent off the NASDAQ. This spooked individual and institutional investors alike, igniting a sell-off that in just six days dropped the NASDAQ by almost 9 percent. By 2001 the bubble was losing air quickly. Most dot-coms had ceased trading after burning through their start-up capital, many having never earned a profit and most losing millions of dollars and leaving investors with a hangover that would last for years. Many businesses and investors would never recover.[52]

At the time of this writing, the NASDAQ sits at 2,130—approximately 40 percent of its closing on March 10, 2000. That means that many of the hoi polloi, Americans who had taken only a passing interest in stocks before becoming passionate investors during the boom, are down 60 percent as well. In the glory days of the dot-coms, any stock with an *E* in front of its name was fair game—a practice some called "prefix investing." Be honest now, did any of you dear readers invest in eToys? The ill-fated Internet toy retailer reached $80 during its May 1999 IPO but had dropped to less than $1 when it declared bankruptcy in February 2001—less than two years later. The failed global online fashion retailer Boo.com spent $188 million in just six months but to no avail. The company closed its doors in May 2000. I could go on listing corporate examples for a long time.[53]

Same for individual stories: I could tell you of college professors, accountants, chemists, and grocery baggers all scarred by the demise of dot-com stocks. One such person, let's call him Fred, was a respected academician. After suffering huge losses when his technology stock-play went horribly wrong, he is now living out his sunset years in a one-bedroom apartment. At seventy-two, he works a night job to make ends meet. Or meet James, the accountant (near retirement) who was so heavily leveraged in technology stocks that when the bubble burst he soon filed for bankruptcy. A short time after that he passed away, his now destitute wife commenting that the stock market crash had crushed his spirit. Regrettably, the memory of that once proud and hard-working man is now sullied by a last-gasp effort to achieve the American Dream.

Let's go back, for a moment, to our comparison of the Gold Rush and the dot-com era. As noted above, both events fostered the notion that the American Dream was more about quick riches and luck than about hard work, sacrifice, and thrift. That shift in perspective put us on track for further problems. The bursting of the dot-com bubble led directly to the housing bubble and bust, our next topic of discussion. Investors, burned in the stock market, were looking for a safe place to invest. The speculative frenzy that had led to the dot-com boom now shifted to real estate, with all too predictable results. "The materialistic display of the big house," observes Yale economist Robert Shiller, "became a salve to bruised egos of disappointed stock investors."[54]

The quest for money and financial security has always been an integral part of the American Dream. As important as the actual content of the dream (the exact amount of wealth, for example) is the means by which we go about attempting to achieve it. Instead of working hard and slowly building a nest egg, too many Americans are setting their sights on what appears to be "easy money," what historian Jim Cullen refers to as "effort-less wealth."[55] And it's not simply economic independence and security we're after, but extravagant wealth.

As professor Matthew Warshauer observes, "It seems that many Americans covet the easy road to the dream and in the process undercut the core values that established the dream in the first place."[56] The hard work and sacrifice that were part and parcel of the original dream have been replaced by wishful thinking of material success, with no willingness to pay the dues necessary to create such wealth. We want to enjoy the trappings of wealth and "look the part," but we don't like the sustained effort and sacrifice.

This is true even in today's frigid economic climate. As we see our hopes of achieving the American Dream slip from our grasp, we pine for "easy money" and chase the many rabbits that promise us effortless wealth. Such desires are at the very core of what makes us human. As Cullen wisely puts it, "A longing for a life of leisure has virtually universal appeal, and given the grind of exertion and duty that has characterized everyday existence for most of human history, it's not hard to understand why."[57]

THE SUBPRIME AMERICAN DREAM

The current housing crisis is Exhibit A in just how far we have strayed on the road to the American Dream. You won't find a stauncher proponent of homeownership than I am, but we've paid too high a price. We've made a colossal mess of one of the most sacred elements of the American Dream: homeownership. Let me explain, as briefly as I can, how something so cherished and seemingly innocent as owning one's own home brought the U.S. financial system to the brink of collapse.

In the table below, I show the major legislation used by the U.S. government to encourage homeownership. It is truly a legacy of shame, though it may be a case of good intentions gone awry. As the table attests, the government attempted (albeit misguidedly) to make the American Dream of homeownership a reality for all Americans. In and of itself, this doesn't sound like such a bad thing. But perhaps not every American deserves to own his or her own home. (Feel free to substitute "can afford" for "deserves" in the preceding sentence.) This is not a racial issue, though minority and majority populations are affected unequally; rather, it is a financial issue. With approximately two-thirds of all Americans owning their own home in 2008, many (if not most) creditworthy households had already taken the plunge into homeownership. The remaining Americans, for assorted reasons, were just not homeownership material.

Federal Encouragement of the American Dream of Homeownership

YEAR	SELECT LEGISLATION & LOWLIGHTS	PURPOSE
1862	Homestead Act (Abraham Lincoln)	Give 160 acres to homesteaders
1913	Tariff Act	Make home mortgage interest deductible
1922	Own Your Own Home Campaign (introduced by soon-to-be President Herbert Hoover)	Defeat the "menace" of tenancy and encourage homeownership
1932	Federal Home Loan Bank System (President Hoover)	Provide a stable source of funds for banks
1933	Home Owner's Loan Corporation (HOLC), a New Deal bailout program	Buy troubled mortgages from lenders
1934	Federal Housing Administration (FHA)— National Housing Act	Insure mortgages
1938	Federal National Mortgage Association (Fannie Mae)	Purchase insured mortgages
1944	GI Bill	Grant government-subsidized mortgages to returning veterans
1949	Housing Act	Expand the federal role in mortgage insurance

YEAR	SELECT LEGISLATION & LOWLIGHTS	PURPOSE
1960s–1970s	FHA's Urban Loan Program	Make home loans to poor African Americans
1977	Community Reinvestment Act (CRA)	Force banks to make home loans to poor neighborhoods
1986	Association of Community Organizations for Reform Now (ACORN)	Campaign to lower lending standards for minorities
1992	Congressional legislation focused on easing credit standards—focus on lending giants Fannie Mae and Freddie Mac	Require Fannie Mae and Freddie Mac to devote 30 percent of their loan purchases to mortgages for low- and moderate-income borrowers
1994	Housing Secretary Henry Cisneros of the Clinton administration declares goal of increasing homeownership	Expand homeownership among lower-income renters
1994	Mortgage Bankers Association (MBA) agreement with HUD	Increase lending to minorities and "rewrite" lending standards
1995	National Homeownership Day, launched by Bill Clinton	Increase homeownership options for all Americans
1999	Revised HUD lending levels for Fannie Mae and Freddie Mac	Increase loans to low-income borrowers by lowering credit standards
2000	American Homeownership and Economic Opportunity Act	Increase homeownership among the "underserved" by lowering lending standards
2003	HOGAR (Hispanic Ownership, a Growing American Reality) launched by the Congressional Hispanic Caucus	Increase homeownership among Hispanics by lowering lending standards
2004	Congressional pressure on Fannie Mae and Freddie Mac	Increase affordable housing by lowering lending standards

Depleting the supply of qualified homebuyers, however, hasn't stopped the U.S. government from trying to increase homeownership, in what

Steven Malanga refers to as the government's "obsessive housing disorder." Malanga, in commenting on many of the legislative acts listed in the table, suggests that "nearly a century of Washington's efforts to promote homeownership has produced one calamity after another. Time to stop."[58] And that long phrase was just the subtitle to his article on Washington's involvement in the homeownership business!

When the housing bubble burst in 2008, critics were quick to seize on the policies of the Bush administration and its efforts to expand homeownership. And they weren't wrong: policies promoted by Bush and his allies pushed for relaxed lending standards (if the word "standards" was still applicable in 2008), encouraged governmental enterprises such as Fannie Mae and Freddie Mac to make mortgages readily available, and all but strong-armed private lenders into being more "innovative"—in other words, lowering standards—so that all Americans could experience the joys of homeownership. The outcome of all this was that millions of people were lured (albeit willingly) into taking on mortgages that they couldn't afford. The subprime mortgage crisis that still reverberates at this writing was the result of banks lending money to people who could not afford it—plain and simple.

But to accuse Bush and his administration of starting this fiasco is to ignore history. As the table above makes clear, a long line of both Democrats and Republicans have had their hands in this mess. Indeed, the American Dream of homeownership is part of the nation's very fabric. In the words of history professor Thomas Sugrue, "To own a home is to be American. To rent is to be something less."[59] As Steven Malanga muses, "The ideal of homeownership has become so sacrosanct, it seems, that we never learn from these disasters. Instead, we clean them up and then—as if under some strange compulsion—set in motion the mechanisms of the next housing catastrophe."[60] Case in point: Obama's recent bailout program, which came to the rescue of many troubled homeowners. Homeowners will likely make the same mistakes again (bought too big a house or tried to "flip" a house as an investment strategy, for example) once the real estate market rebounds.

◆ ◆ ◆

BUT DO WE HAVE TO OWN OUR HOME TO FEEL A STRONG AFFECTION FOR IT?
Being a product of our times, I too have bought into the American Dream
of homeownership. As Sugrue stated earlier, to rent is something less.
Whether we would admit it or not, homeowners look slightly askance at
renters—and, heaven forbid, please don't let them move into our neighborhood and drive down property values with grass up to our knees and
other signs of neglect. For good, and for bad, Americans are obsessed
with owning their own homes.

I am not going to turn this into a manifesto on how the U.S. government became inextricably intertwined in the homeownership business.
Still, I need to describe the highlights (and lowlights) of Washington's
often misguided attempts to bring homeownership to the rank and file of
the American people. Let's go way back to the country's origins.

I think it was humorist Will Rogers who once said, "Buy land; they
ain't making any more of the stuff." Land we had. After we won our independence from the English we became, in the words of historian Jim
Cullen, "a frontier state, in that land became an avowed instrument of
government policy."[61] Thomas Jefferson was an early proponent of homeownership. His republican dream was a vision of independent farmers
owning and occupying the land they worked. Jefferson's work, and that of
many others, held out the promise of a society where the majority of its
citizens owned both a patch of land and the home that sat upon it.

Jefferson's desire to give land away would have to wait until the official
mandate of the 1862 Homestead Act. The Republican Party's victory in
the 1860 election came down to its offer of free land through that proposed act, which was signed into law by President Abraham Lincoln two
years later. Jefferson's plan had come to fruition, as promised. The deal
was a simple one. Any head of household or adult male who was a citizen,
or an immigrant who intended to become a citizen, could claim 160 acres
of land. For this privilege, the prospective landowner paid a small registration fee and promised to stay on the land for five years. After that time,
the deed would be transferred to the homesteader.[62]

Unfortunately, the Homestead Act, like much of the housing legislation to follow, did not live up to its promise. Many homesteaders lacked

the skills and capital necessary to make a go of farming—even on free land. Speculators swept in and purchased most of the land at rock-bottom prices. This sounds eerily similar to today's mortgage crisis, doesn't it? Homesteaders—like today's homeowners—got in over their heads and lost their homes to foreclosure or at drastically reduced prices. Despite its failings, however, the Homestead Act firmly embedded land- and home-ownership as part of the American Dream.

As the decades went by, Americans from all walks of life were committed to the dream of homeownership. Not surprisingly, this penchant for homeownership was particularly keen among the struggling immigrant population. Cullen cites a study showing that by 1900, 55 percent of German, 46 percent of Irish, and 44 percent of Polish immigrants in Detroit owned their own homes. Unlike Europe, America offered unparalleled access to homeownership. Despite what would now be considered un-reachable terms—down payments of 50 percent and five- to seven-year mortgages—40 percent of immigrant families in the United States owned their own home by the late 1920s.[63]

The turn of the twentieth century saw the government take an increasingly aggressive posture on the issue of homeownership. Hoping to guard against what he saw as the rising tide of tenancy, then Secretary of Commerce Herbert Hoover initiated the "Own Your Own Home" campaign. The 1920 census had revealed a small dip in homeownership rates, which stood at 45.9 percent of all households in 1910 but had dropped to 45.6 percent in the intervening decade. Hoover blustered that such a trend spelled doom for the American public. If something wasn't done about it, Hoover warned, we would soon be a nation of renters. He urged banks, builders, and real estate mavens to pitch in to reverse this disturbing development. To an extent, Hoover's campaign worked: the 1930 census found that homeownership had increased to 47.8 percent.

But this was a qualified success. Foreclosures were also climbing, rising from 2 percent in 1922, to 9 percent by 1926, and ultimately 11 percent by 1927. The shared vision that the good times would never end had led many Americans to enter into ill-advised home mortgages. The Great Depression was the exclamation point at the end of this ill-fated experiment. By

1933, nearly 1,000 homes were being foreclosed on a daily basis.[64] By the end of the Depression, notes Malanga, "the U.S. government had become the dominant force in the mortgage market."[65] The Home Owner's Loan Corporation (HOLC) was a New Deal bailout program that bought troubled mortgages from lenders. The establishment of the Federal Housing Administration (FHA) in 1934 and the founding of the Federal National Mortgage Association (later Fannie Mae) in 1938 would both play significant roles some seventy years later in the subprime mortgage crisis of 2008.

The GI Bill of 1944 extended the well-deserved benefits of returning soldiers to include government subsidized mortgages. This had its intended effect, increasing homeownership to over 50 percent of all American families—with 40 percent of the loans subsidized by the government.[66]

In response to mounting political pressure to continue this upward trend, the government loosened its mortgage lending standards to make it possible for less creditworthy families to take the plunge into homeownership. For example, the government reduced the size of required down payments, allowed a higher ratio of house payment to income, and extended the allowable length of mortgages.

As should have been expected, foreclosures on these watered-down government-backed mortgages mushroomed. Failure rates of FHA-insured loans, for example, increased by a factor of five from 1950 to 1960. VA loans doubled over the same period. Meanwhile, the foreclosure rates on conventional loans hardly budged over this time span.[67] For the time being, traditional lenders could stick with the normal (stricter) underwriting standards that had proven themselves as worthy indicators of a homeowner's ability to repay his mortgage. This pattern of increasing homeownership followed by a loosening of credit standards and a spike in foreclosures would play itself out again with the 2008 mortgage crisis.

The legacy of shame regarding Washington's involvement in the home mortgage industry began in earnest with its Urban Loan program in the 1960s and 1970s. The original intent of this program was to quiet the urban unrest in big cities such as Detroit, Los Angeles, and Cleveland. Riots in the mid-1960s had created a tinderbox that could ignite at any time. Politicos from both parties reasoned that extending the American Dream of homeownership to poor blacks and Puerto Rican immigrants, among others, would stabilize the inner cities. FHA-insured loans were given to the poor with as little as $250 down payment. Twenty percent of the mortgages granted in Detroit during this time were granted to single mothers on welfare. These buyers were simply not qualified to be home-owners, and thus foreclosures spread like wildfire in cities across the United States. By the end of the program, the government had lost $1.4 billion in assets across the country.[68] As Steven Malanga concludes, "the meltdown couldn't have happened without Washington's unexamined as-sumption that homeownership would transform the lives of low-income buyers in positive ways."[69] According to Harvard historian Louis Hyman, who compared the scandals associated with the FHA's Urban Loan pro-gram to the current subprime debacle, you can't "solve a problem of wealth creation through debt creation."[70]

Washington would now enlist the help of private lenders in its attempt to boost homeownership. The Community Reinvestment Act (CRA) of 1977 forced banks to make home loans to people in poor neighborhoods. Banks would not be allowed to expand if they didn't make a sufficient number of loans in these neighborhoods. This campaign would last for thirty years and contribute to the avalanche of foreclosures that would ignite a worldwide financial crisis.

Still, many housing advocates felt that mortgage lending was rac-ist—in other words, that lending criteria were too high to allow minori-ties equal access to homeownership. The now infamous activist organization ACORN (Association of Community Organizations for Reform Now) began a campaign in 1986 to lower said lending criteria. Lending standards, argued ACORN, needed to be more flexible for mi-nority borrowers. The most appalling example of such lower standards was the argument that public assistance and food stamps should be con-sidered income. Lower down payments (how much lower could one go?) and an easing of the requirement that borrowers have at least two or three months of mortgage payments in the bank at closing were also part of the mix.[71]

Congressional legislation in 1992 required that 30 percent of all the mortgage loans that Fannie Mae and Freddie Mac purchased be from low- and moderate-income borrowers. With the inevitable increase in foreclosures to follow from the above-mentioned lowered standards, this exposed the mortgage giants to significant risk: one-third of their portfolio now consisted of high-risk loans.[72]

Enter the Clinton administration. In a statement upon signing FHA legislation in 1994, Clinton said, "Homeownership is one of the foundation stones of the American Dream. Renewing and expanding this dream is one of my administration's highest priorities and deepest commitments."[73] Housing Secretary Henry Cisneros made it his stated goal that he would increase homeownership among those inhabiting the lower rungs of the socioeconomic ladder. Of course, this meant lowering lending standards even more—dropping the down payment requirement under certain circumstances and channeling more private money through the CRA and other existing lending laws into low-income lending programs. Fannie Mac and Freddie Mac wasted no time dropping their qualification standards. Bad credit? No problem. Even income from seasonal jobs and public assistance would be considered income. Different decade—same mistakes. The fact that people couldn't afford a home, argued homeownership proponents, didn't mean they should be denied access to the American Dream.

By the late 1990s all hell was about to break loose. Lenders threw caution to the wind and started feeding at the trough of low-income mortgage programs. There was no place to go but down—creditworthiness speaking. In a *Wall Street Journal* article titled "The New American Dream: Renting," Thomas Sugrue had this to say about the looming mortgage crisis: "During the wild late 1990s and the first years of the new century, the dream of homeownership turned hallucinogenic. The home financing industry—at the impetus of the Clinton and Bush Administrations—engaged in the biggest promotion of homeownership in decades."[74] Subprime lending and the securitization of home mortgages, notes Sugrue, "made it possible for more Americans than ever to live the dream or to gamble that someone else would pay them more to make their own dream come true."[75] But like all dreams, eventually you wake up. And what we awoke to was a financial cataclysm to rival the Great Depression. The American Dream had become the American nightmare.

THE CAT'S OUT OF THE (SHOPPING) BAG

> Not everything that can be counted counts and not
> everything that counts can be counted.
>
> **—ALBERT EINSTEIN**

How do you feel about how happy you are? Please answer this question using the following delighted-terrible (D-T) scale:

7	6	5	4	3	2	1
Delighted			Neutral			Terrible

This same question has been asked of many different groups, including forty-nine members of one year's *Forbes* magazine list of wealthiest Americans, with intriguing results. At the time of the survey these wealthiest Americans had a minimum net worth of $125 million and an annual income of at least $10 million. The comparison group of typical Americans had an average family income of $36,000 and a net worth of $122,000.

The two groups (wealthy and nonwealthy) were compared on a number of happiness dimensions, including percent of time spent happy, the delighted-terrible continuum described above, life satisfaction, and levels of positive and negative affect (that is, good and bad feelings). Data was collected via a brief mail survey. Results show that the superrich scored higher on the above delighted-terrible scale (5.87 vs. 5.34) than did the nonwealthy comparison group. The difference, however, was modest: .53 on the seven-point scale. The average for a national random sample in the United States was 5.45—a mere .42 below the wealthy group's average on the same scale. And it gets more interesting. Take a look at several other groups and their average score on the D-T scale:[1]

Life Satisfaction for Various Groups

GROUP	RATING
Forbes magazine's "richest Americans"	5.8
Pennsylvania Amish	5.8
Inughuit (Inuit of northern Greenland)	5.8
African Masai	5.7

What's most intriguing is that although the superrich are pretty happy, the Pennsylvania Amish, the Inuit, and the African Masai are equally satisfied. The Masai, by American standards, live simple lives with few amenities. They live in dung huts with no electricity or running water. Although not nearly as spartan as the Masai, the Amish and Inuit by no means live the life of the rich and famous. A simple conclusion can be drawn from all of this: luxury is not a requirement for well-being.[2]

DESPITE THE NEARLY UNIVERSAL APPEAL OF BEING HAPPY, FEW OF US HAVE discovered the secret of true happiness. As French philosopher Jean Paul Sartre put it nearly fifty years ago, "Everything has been figured out, except how to live."[3] Happiness is an elusive concept: it's hard to put a finger on what it is, but we all know it when we see or feel it. For many of us, happiness is a fleeting emotion attached to a pleasurable experience—that "warm fuzzy" we get when we receive a compliment or find an unexpected five dollars in our pants pocket. But to those of us who have attempted to study happiness, it's something more permanent. Instead of thinking of happiness as an ephemeral state of mind, we see it more as a mood. This is an important distinction, because moods are more stable than emotions. As a mood, happiness is a deep feeling that is always present, even if at times we think it has escaped us.[4]

This more permanent perspective on happiness allows us to place it under our microscopes and take a more careful look at it. And study we have. Hundreds of research projects have attempted to better understand what happiness is, what things make us happy, and how happiness impacts other aspects of our lives.[5] The purpose of this chapter is to explain the nuanced relationship between materialism and happiness. Given today's culture of consumption, and the current economic crisis, I think that relationship is the most important issue facing us as individuals, as a nation, and as part of the larger global community. We are in the midst of what some have referred to as the "Great Disruption"—a time when Mother Nature and Father Greed have hit the wall at once.[6]

I am, however, a cautious optimist. I have said for years that our current obsession with material possessions could be stopped—okay, slowed down—only by an extended economic downturn. Well, we got the downturn, and it appears to be working: people across America, many by choice and others out of necessity, are reconsidering the role money and possessions play in their lives. My hope is that this book will help these same people make the difficult transition away from materialism and that they will come out on the other end happier for their efforts.

When studying happiness, we often use the term "subjective well-being" (SWB) to represent the construct of happiness. SWB is perhaps best understood as the cognitive and affective evaluation of one's own life. The major components of SWB include positive and negative affect and overall satisfaction with one's life, as well as satisfaction with specific life domains. How happy you are has a lot to do with how satisfied you are with important domains such as your family, friends, residence, health, fun and enjoyment, money, and job.

It might be comforting to some and discomforting to others that there is evidence of a biological "set-point" that determines approximately half of one's level of happiness. This set-point varies across individuals, ranging from chronically low levels of happiness, to moderate levels of happiness, to high levels of happiness. A full 90 percent of this set-point is thought to be genetically determined! Although positive or negative events may send people to the lower or upper regions of their set-point, feelings of positivity have been found to be remarkably stable.[7]

Martin Seligman, author of *Authentic Happiness*,[8] has developed what he refers to as the happiness formula:

$$H = S + C + V$$

In this formula, H represents your enduring level of happiness, S is your set-point (or what Seligman calls "set range"), C stands for the circumstances of your life, and V represents factors under your voluntary control.

As mentioned above, approximately half of your happiness is determined by the first element in the equation—that is, by your set-point. Like it or not, we are a function of the gene pool from which we were born. But there's hope: other factors impact your happiness.

The next component of the happiness equation is circumstances. Simply put, the particulars of your life can improve your happiness. The problem is that changing these circumstances can be difficult and expensive. This book's focus is on the relationship between happiness and one important life circumstance: consumption. The importance of money and material possessions in our current consumer culture has given money the power to dramatically sway our happiness.

The last component of Seligman's happiness formula is V which represents factors under your voluntary control. As it relates to money and possessions, what is under your control is the value you place on those material goods relative to the myriad of other things competing for your attention. As we will discuss shortly, how you relate to money and possessions impacts how you feel about yourself, the quality of your personal relationships, and your involvement in the larger community.

Before we have that discussion, you might be interested in measuring your own happiness level and seeing how you stack up against the rest of us. The happiness scale that follows, devised by Sonja Lyubomirsky, will give you a good sense of your level of enduring happiness.[9]

General Happiness Scale

For each of the following statements and/or questions, please circle the point on the scale that you feel is most appropriate in describing you.

1. In general, I consider myself:

   1   2   3   4   5   6   7

   Not a very happy person   A very happy person

2. Compared to most of my peers, I consider myself:

   1   2   3   4   5   6   7

   Less happy   More happy

3. Some people are generally very happy. They enjoy life regardless of what is going on, getting the most out of everything. To what extent does this characterization describe you?

   1   2   3   4   5   6   7

   A great deal   Not at all

4. Some people are generally not very happy. Although they are not depressed, they never seem as happy as they might be. To what extent does this characterization describe you?

   1   2   3   4   5   6   7

   A great deal   Not at all

To calculate your happiness score, add up your score for each question and divide the total by four. The average for US adults is 4.8. Most people score between 3.8 and 5.8.

So, how did you do? Are you one of those terminally happy people that we all wish we could be or, like the rest of us, are you hovering around the average, with a real desire to be happier? If the latter, this book's for you. In our current consumer culture, we imbue money and possessions with almost mystical powers, seeing them as cure-alls for what ails us. My research, however, paints a very different picture.

THE GOOD LIFE

Poll after poll shows that Americans believe, no matter their income level, that they need *more* to live the good life. Even those Americans living at the highest economic range feel that more money and more possessions are needed. A great study from the late 1970s makes this point very well. Researchers asked a large sample of adults to check off which high-ticket consumer items they currently owned or took part in from a list that included things like a car, a swimming pool, foreign travel, and a house. They were also asked, "When you think of the good life—the life you would like to have—which of the things on this list, if any, are part of the good life as far as you are personally concerned?"

As you might expect, the "good life" was always just a bit out of reach. In 1978, the thirty- to forty-year-olds in the sample owned an average 2.5 items from the list but felt that the ideal number was around 4.3 items. In 1994, sixteen years later, the same people were interviewed again. Times had been pretty good to most of those now forty- and fifty-somethings. These fine Americans now owned an average of 3.2 of the big-ticket consumer items. Finally, the good life was within reach. Think again. It appears that the equation for the good life had changed. These same people now felt that 5.4 items were needed to live well![10] A great story, and further proof that our escalating consumption brings us no closer to the so-called good life; it only speeds up the treadmill. (We will take a closer look at the treadmill of consumption in chapter 5.)

This disconnect between increased consumption and happiness is not limited to the United States. European countries such as Belgium, France,

West Germany, Ireland, Italy, the Netherlands, and Great Britain all experienced significant economic growth in the 1970s and 1980s, but they saw little or no increase in their collective well-being during that time. People in Ireland and Great Britain actually saw their life satisfaction drop during the period of 1973 through 1990.[11]

Following World War II, the Japanese economy experienced one of the greatest economic booms the world has seen. During the period of 1950 through 1970, Japan's economy (measured by output per person) grew by more than seven times. During the period of 1958 through 1987, the real income of the typical Japanese citizen increased by a factor of five. In three decades, then, Japan transformed itself from a war-torn country into a world superpower. Sadly, however, this surge in economic output and real income did not result in happier citizens. Researchers discovered that Japanese citizens were no happier; indeed, some polls showed them to be even less satisfied with their lives. As the next figure depicts, Japan's gross domestic product (GDP) increased significantly from 1958 through 1987. However, as the figure's lower line indicates, increased income and spending did little to cheer up the prospects of the typical Japanese citizen. Japan started 1958 in poverty and arrived at 1987 as one of the world's wealthiest nations, yet as a nation its collective well-being was little changed.[12]

This seeming contradiction became the basis for a theory known as the "Easterlin paradox." In 1995, Richard Easterlin, then an

Japanese Happiness and GDP, 1958–1987

economist at the University of Pennsylvania, published a paper which argued that economic growth does not necessarily lead to happier citizens.[13] Easterlin showed that people in poor countries *do* become happier once they can afford basic food, shelter, and clothing—not a big surprise. But what was surprising was Easterlin's finding that, after these basic needs have been met, additional income and possessions add little to one's life satisfaction. Recent research by Ed Diener, a leading happiness researcher, found that positive and negative affect, job satisfaction, and life satisfaction all level off when household income reaches $40,000–$60,000 per year.[14] This is additional evidence that more money does not necessarily mean greater happiness. Further gains appear to simply raise the bar on one's desires—a process commonly referred to as "adaptation." As noted earlier, we humans tend to quickly adapt to changes in our life circumstances, considering our new situation the baseline from which to judge future events.

One contemporary example of the Easterlin paradox is the cell phone. At first it was enough simply to be able to make and receive calls away from home. Now if you don't have e-mail and Internet capability on your little phone, you're practically in the Stone Age and are at risk of social ostracism—cast as a twenty-first-century Luddite.

Another example of the disconnect between income and happiness can be found in modern-day China. Following unprecedented economic growth from 1994 to 2005, during which time real income grew by 2.5 times, a survey of 15,000 individuals conducted by the Gallup Organization found no reported increase in life satisfaction. This growth meant a substantial increase in material well-being, to be sure: color television ownership increased from 40 percent to 82 percent, for example, and telephone ownership rose from 10 percent to 63 percent. Yet all indications were that life satisfaction remained static and, in many instances, even went down with the economic boom. Research by Hilke Brockman and colleagues at Jacobs University in Bremen, Germany, paints an even bleaker picture. Their data suggests that the percentage of Chinese who described themselves as "very happy" dropped like a lead balloon from 28 percent in 1990 to 12 percent by 2000.[15]

TRUE LUXURY   Maybe the Finns can teach us a lesson or two about true luxury. In an article that appeared in the *Christian Science Monitor*, Trevor Corson explains that his trips to Finland changed how

he looked at material possessions. Before his two trips to that country, Corson had defined true luxury as "a flat-screen TV the size of Kansas and a leather-upholstered car that can travel at triple the speed limit." This, however, was about to change. As Corson strolled the streets of Finland's capital, Helsinki, he noticed a decided lack of ostentatious architecture and fancy homes, and the cars he saw tended to be quite modest. All a bit dull for a card-carrying American. On his next visit in July, the city felt like a ghost town. On asking why, he learned that the typical Finn takes a five-week summer vacation, usually to an isolated cabin with few creature comforts or things to do. Corson decided to give it a try: after a couple of days, the author was a convert. The lack of anything to do was relaxing—a form of luxury that is a far cry from our mad dash to acquire and display our possessions. Time and solitude, for many Finns, are to be treasured over material trappings.

Despite the peace and serenity he felt after his last visit to Finland, Corson was happy to be back to the hustle and bustle of New York City and to the optimism and entrepreneurial spirit that make America so great. In referring to the present economic crisis, Corson hit the proverbial nail on the head when he offered this little chestnut: "As we struggle to get back on our feet, perhaps we should pause for our own 'Finnish Moment.'"[16]

MATERIALISM AND INDIVIDUAL HAPPINESS

The study of materialism and its impact on individual well-being is a relatively new field, maybe twenty-five years old. In that time, however, a solid body of research findings has emerged that links materialism to a broad array of negative personal outcomes. Early research by Russell Belk associated materialism with such undesirable traits as envy, greed, and non-generosity and found that these unseemly traits were negatively correlated with one's happiness and life satisfaction.[17]

Subsequent studies in the fields of psychology, economics, and consumer behavior have found that materialism is inversely related to positive affect, happiness, self-esteem, self-actualization, subjective well-being, quality of life, and satisfaction with life. Materialism has also been found to be associated with higher negative affect, increased social anxiety (and anxiety in general), self-criticism, and increased time spent unhappy and/ or depressed. Moreover, materialists report lower levels of physical well-being; they complain more about backaches, bumps, and bruises and report more headaches, colds, and bouts of flu than their less materialistic counterparts.[18]

IN THE CONSUMER REALM, MOST RESEARCH HAS SHOWN THAT MATERIALISM wreaks havoc on personal finances, is a major contributor to credit card abuse, and often goes hand in hand with compulsive buying. Some studies have found that materialism is negatively correlated with satisfaction with one's finances and standard of living and is positively correlated with the amount of TV a person watches. What is clear from all of these studies is that those of us who place a high value on material possessions consistently report lower levels of well-being than those who place less importance on such values.

In 2006, my colleague Aimee Clement and I conducted a large-scale survey of adults from across the United States and from all walks of life. Our objective was to assess the relationship between the respondents' level of materialism and their satisfaction with the quality of their life. Quality of life was measured by asking respondents how satisfied they were with eight individual life domains, including family, friends, residence, health, amount of fun and enjoyment, money, and job. Materialism was measured using the scale found in the Appendix, and a five-point scale (ranging from very dissatisfied to very satisfied) was used to record responses.

The study's results were quite clear: quality of life and materialism were negatively correlated. This means that the more materialistic respondents were less satisfied with their lives than respondents who were less materialistic. In summary, materialists were less satisfied with their family, their friends, their self-perception, where they lived, their health, the amount of fun and enjoyment they experienced, the money they made, and their jobs.[19]

A series of research studies by Emily Solberg, Ed Diener, and Michael Robinson is a good complement to my quality of life study with Aimee Clement.[20] These investigators hypothesized that the negative relationship between materialism and well-being exists because "people generally are more distant from their material goals than from most other goals." A possible reason for this is because TV (which Americans watch a lot of—four to five hours every day) and advertising constantly bombard us with images of tempting goodies that we currently lack. Other research has confirmed the TV linkage: people who watch more TV do in fact have a larger gap between their current and ideal states than those who watch less TV.[21]

Solberg and company decided to test their "gap" hypothesis with several very large samples of college students. Why college students? Because they have little income while in school and yet likely have high material aspirations because of their earning potential after graduation. All in all, college students are an excellent population in which to test the hypothesis that the wider the gap between present status and ideal status, the greater the negative relationship between materialism and well-being.

Their first study used a sample of 13,500 students from thirty-one countries around the world. Students rated their satisfaction (on a nine-point scale) across a number of life domains, including finances, friends, family, work, religion, health, significant others, self-esteem, and fun—a lot of the same life domains covered in my earlier research with Aimee Clement. Results showed that the largest gap between current and ideal state was in the financial domain. By and large, students were not happy with their financial situation. Because of this wide financial gap, students who place a high value on money and material possessions may be less happy than those who are less materialistic; the latter group places a higher value on other life domains where the gap is narrower, thus promoting well-being.

In a second study of over 7,000 college students from forty-one countries, the investigators had students report (using a seven-point scale) how satisfied they were with various life domains. As you might have guessed, they were least satisfied with their finances. The table below shows the average satisfaction level for selected domains, in declining order.

The results from these first two studies confirm the importance of money and possessions in determining individual well-being. Students from around the world were least satisfied, in assessing their various life domains, with their financial situation, despite the fact that most of them came from affluent societies. The open-ended nature of material goals and a consumer culture that bombards us with commercial messages of the good life through consumption make it unlikely that any of us will ever feel that we have achieved, in the words of fiscal guru Dave Ramsey, "financial peace."

As the authors point out, this distinction does not hold for many

Student Satisfaction with Life Domains

1. Friends—5.45
2. Family—5.38
3. Freedom—5.08
4. Education—4.89
5. Self—4.83
6. Religion—4.79
7. Nation—4.40
8. Romance—4.35
9. Travel and Textbooks (tied)—4.20
10. Finances—4.11

HIGHER SCORES = MORE SATISFIED

of the other nonfinancial domains, because our goals for those domains have inherent limits. Most of us need only one spouse, for example; we can get only so much education; and while we may want to be healthy, we'll settle for manageable cholesterol levels and an absence of disease. Yet the sky's the limit when it comes to money and material possessions. *That's* why the financial gap is likely wider than the gap in other life domains.

Love, Money, and Life Satisfaction

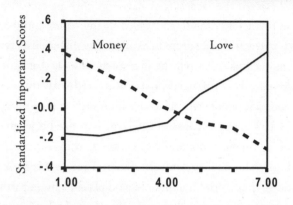

Importance values are standardized within nations, and thus societal differences are controlled.

The darker side of material values was evident in this global sample of young people. Respondents were asked to rate the importance of a list of values including love and money. As can be seen in the figure above, placing a high value on money was inversely related with life satisfaction. In contrast, those who placed a higher importance on love were found to be more satisfied with life.

In a third study in the Solberg series, the investigators directly tested the argument that materialists are further from their material goals and therefore less satisfied than those who place less value on such goals. A sample of sixty-nine students were asked to consider different life goals, including financial success, social recognition, and being physically attractive, as well as feeling good about who you are, having strong relationships with friends and family, and being involved in the community. These goals were part of an "Aspiration Index" developed by Tim Kasser and Richard Ryan that will be discussed at length in chapter 7.

Participants were asked to rate (1) how important success in this area was to them, (2) where they currently were in each area, (3) where the average person their age was in each area, (4) where they currently wanted to be in each area, and (5) where they would ideally want to be in the future in each area.

As hypothesized, the largest gap between current and ideal states occurred in the financial domain (addressed by questions 2 and 5). But could that wide gap in the financial domain explain why materialists tend to be less happy? To answer this question, the investigators looked at how the importance people placed on financial goals interacted with financial gap size. They reasoned that people who placed a high importance on financial success would be less happy because of the large gap size between current and ideal states in the financial domain. Sure enough, using a research technique known as "regression analysis" (where material values were used to predict life satisfaction), they found that people who placed a higher value on material possessions were less satisfied with life when their gap size was large.

This study's results are intriguing. Having materialistic goals per se does not cause unhappiness—provided that you are satisfied with your current financial situation. Likewise, a large gap between your current and ideal financial states will not on its own cause you to be unhappy—provided that you place little value on this gap. The double whammy comes when you place a high value on material possessions and have a large gap in this area: *then* you will be less happy.

MATERIALISM AND MENTAL HEALTH

Evidence of the disconnect between materialism and happiness becomes even more apparent when measures of "ill-being" are considered. Ample evidence supports the claim that as societies become wealthier, they experience an increase in mental health disorders and other related social problems, as well as a plateauing of life satisfaction. While happiness has flatlined in the United States, as we saw in chapter 1, the incidence of depression has skyrocketed. Over the past fifty years depression has increased by as much as tenfold in this country. The typical adolescent in the 1980s seeking treatment reported higher levels of anxiety than his or her counterpart seeking treatment in the 1950s.[22]

More recent studies from Harvard University have found that one in fifteen Americans experiences an episode of major depression each year. Data from the Centers for Disease Control (CDC) shows that the incidence of "mentally unhealthy days" is on the rise. Since 1993, the CDC has interviewed over a million people and asked them, "Thinking about your mental health, which includes stress, depression, and problems with emotions, for how many days during the past 30 days was your mental health not good?" The results of this large survey are not encouraging. In 1993, the average number of "mentally unhealthy days" averaged about 2.9 out of 30. In 2004, the average number of unhealthy days had increased to 3.5—a 21 percent increase.[23] A just released study of nearly 80,000 high school and college students found teens and young adults today are 5 to 8 times more likely to report being depressed or suffering other mental health problems as compared to their counterparts in 1938. The authors of the study conclude this is the result of higher levels of materialism among young people today. And, that people who are more materialistic commonly report higher levels of anxiety and depression.[24]

Large, nationally representative samples of countries from around the world paint a similar picture. The World Health Organization's (WHO) surveys show that the United States is at the top of the heap when it comes to mental suffering. As can be seen below, 26.4 percent of Americans surveyed reported suffering from depression, anxiety, substance abuse, or impulsivity-aggression in the past year.[25]

Mental Illness Around the World

NATION	PERCENT SUFFERING	NATION	PERCENT SUFFERING
United States	26.4	Ukraine	20.5
Great Britain	23.0	France	18.4
Australia	23.0	Colombia	17.8
New Zealand	20.7	Lebanon	16.9

NATION	PERCENT SUFFERING	NATION	PERCENT SUFFERING
Netherlands	14.9	China (Beijing)	9.1
Mexico	12.2	Japan	8.8
Belgium	12.0	Italy	8.2
Spain	9.2	Nigeria	4.7
Germany	9.1	China (Shanghai)	4.3

Clinical psychologist Oliver James attributes much of this suffering to what he calls "selfish capitalism" (SC). An important element of SC is the belief that "consumption and market choices can meet human needs of almost every kind."[26] Based upon the prevalence of this materialistic element and several others, James classifies the English-speaking nations of the United States, Great Britain, Australia, Canada, and New Zealand as having SC economies.

According to James, SC economies share the following characteristics (not held by unselfish capitalist societies):

1. Larger disparities in wealth between the rich and poor

2. Higher proportion of workers earning less than the average wage

3. Wealth in the hands of a small number of wealthy elites

4. Higher personal debt

5. Higher credit card ownership

6. Lower personal savings

7. Longer work weeks

8. Less economic security

James's view that materialism is the culprit in much of today's mental illness jibes with numerous studies by Tim Kasser and others that have found materialistic people to be more likely to suffer depression, anxiety, and substance abuse disorders.[27] Selfish capitalism theorists argue that people in materialistic societies, though rich in possessions, are not getting their fundamental needs for security, autonomy, competence, and community involvement met, and this failure results in higher levels of stress and mental illness. The above list of nations shows that the prevalence of mental illness is indeed higher in English-speaking nations compared to mainland Europe and Japan—in many instances twice as high—supporting James's view that these are the SC economies.

Furthermore, the prevalence of mental illness is on the increase in these SC nations: since the 1970s, they have become more SC in their orientation, while mental illness has also been on the rise. In response to the question, "Have you ever felt that you were going to have a nervous breakdown?" Americans from the 1950s to the 1990s increasingly answered yes; in fact, 15 percent more Americans in 1976 than in 1957 said yes to this question. By the time the survey was conducted again in 1996, the proportion who said yes was 66 percent higher than in 1976. In Britain, similar figures emerged. Surveys of Brits in their thirties or early forties in the years 1946, 1958, and 1970 revealed that reports of mental illness had nearly doubled. Even more alarming were two surveys conducted in Australia in 1997 and 2001—a period, according to James, of accelerated SC governance. The number of Australians who reported psychological distress increased by a whopping 66 percent in four short years.[28]

The contrast between these results and the messages we receive every day from advertisements, peers, and even our government—that shopping makes us happier, more productive citizens—is pretty striking. But *why* are highly materialistic people less satisfied with their lives than their less materialistic counterparts? This all-important question will be answered in the next four chapters, which address the topics of the treadmill of consumption (chapter 5); the financial foibles of materialists (chapter 6); life goals, materialism, and happiness (chapter 7); and personal relationships, materialism, and happiness (chapter 8).

THE TREADMILL
OF CONSUMPTION

Oh Lord, won't you buy me a Mercedes Benz? My friends
all drive Porsches, I must make amends.

—JANIS JOPLIN

Using material possessions to exhibit status is commonplace in today's consumer culture. We may not know our neighbors, but we feel compelled to make sure they know that we're people of value. As humans we rely on visual cues such as material possessions to convey our status to others and to ascertain the status of people we don't know. The quest for status symbols influences both kids and adults, although the objects we choose to display may differ with age. (Cell phones may be an exception that spans all age groups.)

For young people, cell phones are seen as necessities, not luxuries. A teen or even preteen without a cell phone feels set apart, on the outside looking in. This is in part because cell phones are a way to stay tightly connected with others (text messaging "blind," with cell phone in the pocket, is one of my favorites); however, cell phones are also important fashion statements and social props. For young people, cell phones are second only to cars as symbols of independence. Many teens see cell phones as an extension of their personality, and phone manufacturers and service providers, knowing this, give them many options to express their inner selves—ways to personalize their ringtone, change their "wallpaper," and customize their "skin," for example, as well as add many apps and accessories.

Adults, especially men, are also susceptible to the status appeal of cell phones. Researchers in the United Kingdom studied the use of cell phones after reading newspaper stories about nightclubs in South America that required patrons to check their phones at the door. Club managers found, the stories reported, that many checked phones were props—not working cell phones. To learn more about whether and how people were using their cell phones as social props, the researchers studied cell phone use in upscale pubs in the UK. What they found is most interesting: men and women used their cell phones in different manners. While women would leave their phone in their purse until they needed it, men were more likely to take their phone out of their pocket or briefcase and place it on the counter or table in view of all. Furthermore, like peacocks strutting with their plumage in full display to attract a mate, men spent more time tinkering with and displaying their phone when the number of men relative to women in the pub increased.[1]

As LONG AS CONSUMERS ATTEMPT TO SIGNAL THEIR SOCIAL POWER THROUGH conspicuous consumption, the levels required to make a visible statement of power will continue to rise. If person A buys a new car, person B has to buy a better car to compete; and then person A has to buy a boat as well—and so on. But, as we saw in the previous chapter, once basic needs are met there's no additional happiness with additional purchases. The process of moving ahead materially without any real gain in satisfaction is often called "the treadmill of consumption." That treadmill is a barrier to raising your level of happiness, because it causes you to quickly adapt to good things by taking them for granted.

Research has shown that humans are very flexible. We tend to get used to new circumstances in our lives—including financial circumstances, both good and bad—and we make such mental shifts quickly. Economic gains or losses do give us pleasure or pain, but the effects wear off quickly. When our situation improves, having more money or possessions almost instantaneously becomes the new "normal." As our store of material possessions grows, so do our expectations.

Many researchers have likened this process to drug addiction, where the addict continually needs more and more of the drug of choice to achieve an equivalent "high." This means that acquiring more possessions doesn't take us any closer to happiness; it just speeds up the treadmill. I regret to say that there is a great deal of evidence supporting the existence—and potential harm—of the treadmill of consumption.

If the treadmill didn't exist, people with more possessions would be happier than those "less fortunate" souls who own less. But, as we have seen, this simply isn't the case. The "less fortunate" are, for the most part, just as happy as those with more stuff. Big purchases and the piling up of material possessions hold little sway over happiness. Probably the most discouraging proof for this statement can be found in the study of lottery winners. An integral component of the shiny-objects ethos is quick riches. What better way to catapult yourself past your neighbors than to strike it rich with the lottery, right? If you foresee nothing but a lifetime of fun and sun for lottery winners, you're wrong. A study of twenty winners found that they were no happier a few years after their good fortune; in fact, some were even less happy than before they bought their winning ticket.[2] If the lottery can't pull us out of our current torpor, what hope is there for a raise at work, a flat-screen (plasma) television, an iPhone, or a new car (surely the new Lexus would be an exception)?

CONSUMING FOR STATUS

One important reason that consumers buy products is to satisfy social needs. Many of us spend a large proportion of our disposable income on so-called status items, and this trend is on the rise as we continue to embrace the shiny-objects ethos. "Wait a minute," some of you might be saying; "hasn't the current economic crisis stemmed the tide of status consumption?"

My response to that question is that it never has in the past. Sure, we might mind our financial p's and q's during the actual crisis, but we have always returned to our profligate ways once we've navigated our way through the economic doldrums.

You need look no further back than the early 2000s, when the Internet bubble burst and the stock market tanked. It wasn't long until our spending picked up again, and with a renewed vengeance. That's precisely what brought us where we are today. Similar economic corrections in the 1970s, '80s, and '90s produced the same results: we tightened our financial belts only to loosen them when the clouds receded. It's really a lot like yo-yo dieting. Each time after we fall off the financial wagon we're a little worse off than the time before. Apparently as consumers we tend to suffer from short-term memory loss!

Pursuing materialistic ideals is a competitive and comparative process—hence the expression "keeping up with the Joneses." And today, with daily twenty-four/seven media coverage of the lifestyles of the rich and famous, our competition is no longer limited to our neighborhood. Bill and Melinda Gates and the sultan of Brunei have replaced Joe and Irma down the street as our points of reference. To achieve a position of social power or status, one must exceed this expanding community norm. Even the superrich aren't happy. There's always someone with a bigger home or fancier yacht—or, heaven forbid, a prettier wife. Yes, we even use other humans as chattel in our attempt to secure our position in the social hierarchy! The result of all this social posturing is no end to our wants and little improvement in our satisfaction, despite an ever-increasing consumption of goods. And Madison Avenue knows it: after price, status is the principal theme of most advertising.

CAN YOU HEAR ME NOW?   I thought I had found the ultimate status symbol when I came across Motorola's new $2,000 Aura cell phone. The avant-garde Aura sports 700-plus individual components, a stainless-steel housing, and a front plate that takes the manufacturer a month to create. Add to this list the world's first handset with a circular display (great color and resolution!), a sixty-two-carat sapphire crystal lens, a multimedia player, stereo Bluetooth, and much, much more.

!! CONSUMERS !!

!! GONE WILD !!

My amazement over Motorola's Aura was short-lived, however. I lost interest when I heard about the $1 million—yes, $1 million—cell phone from GoldVish (a Swiss company). The phone is made of eighteen-carat white gold and is covered with diamonds. Bluetooth? Of course. How about a two-gigabyte memory, eight-megapixel camera, MP3 player, worldwide FM radio, and e-mail access? Not to worry if a million is a bit rich for you: Gold-Vish has made available several other phones for around $25,000—no doubt delivered in plain brown-paper packaging to avoid any embarrassment associated with buying a cheaper model.[3]

Status consumption has been defined as "the motivational process by which individuals strive to improve their social standing through conspicuous consumption of consumer products that confer or symbolize status to the individual and to surrounding significant others."[4] It is our attempt as consumers to gain the respect, consideration, and envy from those around us. Status consumption is the heart and soul of the consumer culture, which revolves around our attempts to signal our comparative degree of social power through conspicuous consumption. If you don't buy into status consumption yourself, you certainly know people who do. They go by many names, but "social climbers" and "status seekers" will do for now. Climbers and seekers work to surround themselves with visible evidence of the superior rank they claim or aspire to. Most of us, to some degree, are concerned with our social status, and we try to make sure others are aware of it as well.

Status consumption began in the United States as a way for members of the upper crust to flaunt their wealth to each other. Over the past century the practice has trickled down to the lower rungs of the economic ladder. People are willing to go into debt to buy certain products and brands—let's say a $2,500 Jimmy Choo handbag—because these status symbols represent power in our consumer culture. Cars, for example, are an expensive but easy way to tell the world you've made it; there's no mistaking which are the most expensive. The problem is that nearly everyone else is upgrading to the latest model as well, so no real increase in status occurs—another example of the treadmill of consumption. Fortunately—note the irony there—our consumer culture, with its vast array of products, allows us many other opportunities to confer status upon ourselves. Media mogul Ted Turner put it this way: "Life is a game. Money is how we keep score."[5]

Status consumers are willing to pay premium prices for products that are perceived to convey status and prestige. A high-end Patek Philippe watch is a good example of a product that is—and is blatantly marketed as—a quintessential status symbol. One of Patek's advertising slogans is, "You never really own a Patek Philippe. You merely look after it for the next generation." Trust me; you're buying it for yourself. Despite the manufacturer's claims to the contrary, a Patek Philippe does not keep better time than the myriad of cheaper alternatives on the market; on the contrary, it serves primarily as an unambiguous symbol of status. To many people, owning a Patek signals that you've made it. To me, however, it sends the signal that you've forgone a golden opportunity to do good with the money spent so lavishly on a very expensive watch. It's a zero-sum game no matter how much money you make.

And, of course, Patek Philippe watches are only one of a myriad of examples I could use to document our preoccupation with status consumption. What about Lucky Jeans, bling (it's shiny), Hummer automobiles (maybe one of the more blatant cries for help), iPhones, fifty-two-inch plasma TVs, $3,000 Chihuahua lap dogs (think Paris Hilton), McMansions, expensive rims for your car tires, anything couture, Gulfstream jets, Abercrombie & Fitch and Hollister clothes (for teens and preteens)—even drinking water! No consumer product category has been left untouched. Even the most banal, everyday products have been branded—think $2,000 fountain pens.

Today, status is conveyed more often through ownership of status products than through personal, occupational, or family reputation. This is particularly true in large, impersonal metropolitan areas, where people can no longer depend on their behavior or reputation to convey their status and position in society.

MEASURING STATUS CONSUMPTION

Are you trying to keep up with the Joneses—or worse, the Gateses? Let's find out! Take this commonly used scale to measure status consumption.[6] Be brutally honest in how you answer each question. The results won't be truthful unless *you* are.

Status Consumption Scale

Circle the extent to which you agree or disagree with each statement. There are no right or wrong answers, just the truth as you perceive it.

	Strongly Agree	Agree	Neutral	Disagree	Strongly Disagree
1. I am interested in new products with status.	5	4	3	2	1
2. I would buy a product just because it has status.	5	4	3	2	1
3. I would pay more for a product if it had status.	5	4	3	2	1
4. The status of a product is irrelevant to me	5	4	3	2	1
5. A product is more valuable to me if it has some snob appeal.	5	4	3	2	1

SCORING INSTRUCTIONS

1. *Add up your scores for items 1–3 and 5. Your subtotal:* _____

2. *Reverse your score for item 4. That is, if you scored a 5, give yourself a 1; if you scored a 4, give yourself a 2; etc. Your subtotal:* _____

3. *Add your subtotals from steps 1 and 2. This total, which should range between 5 and 25, represents your status consumption score.*

Write your status consumption score here: _____

Based upon a recent sample of 403 adults from across the United States (ranging in age from eighteen to sixty-five) who completed this status consumption scale as part of a larger survey, here are the benchmarks for seeing how you stack up.

If you scored below 16 you're in the bottom half of the country in terms of status consumption (the average being 15.4). If you scored less than 12.5, you're in the bottom 25 percent of the sample (which was chosen to mirror the U.S. population). On the flip side, if you scored above 20 you're in the top 25 percent of U.S. adults. Don't forget, however, that the above results are from a sample of U.S. consumers, who as a group are high-status consumers. Your score might be relatively much higher in a sample from outside this country.

THE DARK SIDE OF SHOPPING

As we saw in the previous section, status consumption creates the perception of social power in the buyer. Because status consumption is by nature competitive, consumers must continually acquire new and more material possessions to send conspicuous signals of wealth and power. This escalation has the potential to lead to a condition known as "compulsive buying."

Compulsive buyers are preoccupied with the importance of money as a solution to problems and as a means of comparison. Like status consumers, they make purchases in an attempt to bring into balance the discrepancy between their identity and the lifestyle projected by various products. In other words, compulsive buying helps them convey a particular, socially desirable appearance. In compulsive buyers the importance of acquisition goes deeper than that, however: compulsive buying is best understood as "chronic, repetitive purchasing that is a primary response to negative events or feelings, and whose primary goal is mood repair."[7]

I have conducted numerous surveys in the United States and Mexico that have found a positive relationship between status consumption and compulsive buying.[8] In these surveys, consumers who scored high on the status consumption scale you just took were more likely than low scorers to be classified as compulsive buyers. Several additional surveys I conducted with members of Generation X (the Baby Bust generation born 1965 to 1980), and the Millennials (young adults born in the 1980s or early '90s) suggest that the incidence of compulsive buying may be increasing with each successive generation.[9]

When you really think about it, we're *all* compulsive buyers. Outside of our very basic needs for food on our plates, a roof over our heads, and clothes on our backs, everything else we purchase is discretionary. But as compulsive buying becomes more severe in an individual, and more prevalent in our society, it causes serious personal, interpersonal, and social problems. My research found, for example, that college students who were classified as compulsive buyers worked longer hours to pay their bills, missed class more frequently, earned lower grades, were more likely to drop out of school, and expressed higher levels of anxiety and depression; they were also more likely, in extreme cases, to commit suicide. Bad debt, revealed by credit checks, kept many students from getting a good job upon graduation, being accepted to graduate school, or getting a home loan.[10]

Let's take a closer look at compulsive buying. First off, it's a chronic tendency to purchase products far in excess of one's needs and resources.[11] Compulsive buying is often triggered by negative events or feelings like a bad day at work or chronic feelings of personal inadequacy—"I'm not attractive", "I don't have any friends", or any of the other myriad of insecurities we may possess as part of the human species. Shopping may quell these negative feelings, but only temporarily. Guilt, shame, and anxiety often follow closely on the heels of a shopping binge. Compulsive buying can become an addiction when it becomes one's primary coping response to negative events and feelings and is performed despite the numerous negative consequences of such behavior. Like substance abusers or behavioral addicts (sex, cellphones, computers, etc.), compulsive buyers lack control over their impulses and often deny the negative fallout of their over-shopping. We often jokingly refer to compulsive buyers as "shopaholics," but the condition isn't funny. The high from shopping can cause reactions in the brain similar to that caused by drugs and alcohol. Compulsive buyers use shopping to escape uncomfortable feelings the same way other addicts use sex, drugs, alcohol, or any combination thereof.

Compulsive buying has both cognitive and behavioral components, and each causes problems. Cognitively, impairment may be manifested through anxiety, feelings of guilt, the fear of being discovered, and troubled interpersonal relationships with family and friends; depression is also common. Research has revealed that compulsive buyers are also more likely to suffer from anxiety disorders, eating disorders, substance abuse disorders, and a host of impulse control disorders. Research that I conducted with Jeff Tanner at Baylor University found that adolescents between the ages of twelve and nineteen who were classified as compulsive buyers were also more likely to smoke cigarettes, drink alcohol, use drugs, and engage in premarital sex.[12]

The behavioral component of compulsive buying often results in financial problems, including credit card debt and bankruptcy. In a large survey of U.S. adults, I found that credit card misuse was a strong predictor of compulsive buying. However, in another study I conducted with Eli Jones, we found that credit card misuse changes the relationship between one's attitudes toward money and compulsive buying. In other words, this study found that even if people held money attitudes that could lead to compulsive buying, they were less likely to do so if they used credit cards responsibly.[13] Credit card debt is a significant portion of debt for many individuals seeking protection under the bankruptcy code—often as much as twice the individual's yearly income. Especially troubling is the fact that compulsive buying can be passed on from parents to their children, making its growth in society a concern. Whether this transmission is via genetics or environment is open to speculation. Like most things in life, it's probably a bit of both nature and nurture. (We'll talk more about the subject of the genetic roots of materialism and shopping in chapter 9.)

Compulsive buying affects a significant portion of U.S. consumers. This may be particularly true for females, who early research suggests comprise between 80 and 95 percent of all compulsive buyers.[14] A study conducted in 2006 by Lorrin Koran and several coauthors calls that long-held assumption into question, however: in a large random sample of 2,500 adults, Koran and his colleagues found that the incidence of compulsive buying is approximately equal across the sexes (females = 6 percent, males = 5.5 percent). The large and random nature of their sample gives pause to the long-held notion of compulsive buyers being largely female. The study's results are intriguing and suggest that the issue of gender and compulsive buying merits continued scrutiny.[15]

Regardless of their representation among compulsive buyers, women account for approximately 80 percent of all consumer spending in the United States. Even in our enlightened age, shopping is still seen as "women's work." Women have been socialized to derive pleasure from shopping; therefore it has become a convenient way for them to express themselves. Through shopping, women can create a desired identity. Much of what is purchased revolves around their appearance: clothing, jewelry, cosmetics and the like. Material possessions are typically seen by women as extensions of who they are or who they would like to be. This broader self-image, including use of products to define who we are, is referred to as the "extended self," as we saw earlier. If our self-definition is extended to the products we purchase and display, the meanings imbued in these products likely reflect important social roles. Gender is one such social role. In the United States, women have been socialized to develop a more passive coping style when dealing with stress. Shopping is something they already do and is a socially acceptable activity (even desirable if you're a "bargain hunter") that can be used as a means of expressing oneself, providing for others, and coping with the headaches of everyday life.

Someone need not be a compulsive buyer to be affected by this disorder. It's not just somebody else's problem. Debt and its attendant life disruption are "defining indicators" of compulsive buying. Burdensome levels of debt, particularly installment loans, credit card debt and large home mortgages can have potentially devastating effects for individual consumers, their loved ones and for economies and societies. From a societal standpoint, out of control spending can lead to higher levels of loan delinquencies and default (think mortgage crisis), tighter credit standards (think credit cards, small businesses loans, and mortgages), less available credit, higher unemployment and interest rates and environmental blight from an increased use of energy and natural resources devoted to the production, consumption, and disposal of an ever-increasing amount of consumer goods. At the individual level, compulsive spending can lead to home foreclosures, debt collection harassment, repossessions, personal bankruptcies, loan defaults, arguments with spouses, children and other family members and friends, and even divorce, as well as the previously noted depression, anxiety, and low self-esteem.[16] Research tells us that highly materialistic spouses are 40 percent more likely than their less materialistic counterparts to experience financial problems, which inevitably leads to lower levels of marital satisfaction.[17] (Chapter 7 will look at the devastating impact of materialism on relationships.)

9/11 SPENDING ORGY Despite having systematically studied consumer behavior for over 100 years, we still don't fully understand why people buy things. As I tell my students at the start of each semester, as consumers we do strange things for even stranger reasons. Take Kathy Trant, for instance. She is the widow

of Dan Trant, once a trader with Cantor Fitzgerald on the 104th floor of the World Trade Center, who tragically died in the 9/11 terrorist attacks. When her husband died, relatives, friends, and complete strangers donated approximately $3 million to Trant and her three children. Based upon her deceased husband's estimated future earnings, the Federal Victim Compensation Fund awarded Kathy an additional $4.2 million, of which she received approximately half. In total, Trant received about $5 million—money she later referred to as "blood money."

Less than four years after the tragedy of 9/11, Trant had spent nearly all of the money in an orgy of frenzied consumerism—$5,000 purses and designer gowns and $800 shoes, closets full of them. Trant turned her home on Long Island into a $2 million estate (tripling its size), traveled the world, and squandered $500,000 on shoes alone. She also spent lavishly on family, friends, and even strangers—$15,000 to her housekeeper to buy a home in her native El Salvador, $4,000 to a total stranger for breast implants, $70,000 to take a group of friends to the Super Bowl, and another $30,000 for a trip to the Bahamas. Trant claimed she wore about 10 percent of the clothes and gave away the majority to friends. This lack of concern about the possessions themselves is typical of many compulsive buyers. It is the act of buying that gives them relief, albeit temporarily, from the pain of their loss.[18]

Compulsive buyers have negative feelings about themselves and their lives in general. Much of their buying is an attempt to escape an awareness of those feelings. Something called "escape theory" attempts to explain why some consumers engage in this rather self-destructive behavior. Escape theory argues that people in whom self-awareness is painful enough to be harmful focus their attention on a single element in their environment. By becoming absorbed in the here and now, they can block out negative thoughts and feelings about themselves. For compulsive buyers, shopping becomes so all-consuming that it overrides the long-term negative consequences.[19]

ARE YOU A COMPULSIVE BUYER?

You probably already suspect the answer to the above question, even before responding to the compulsivity scale provided here. Let me warn you, though: this is a scale used to classify people as *clinical* compulsive buyers—the most extreme cases. If you meet this scale's standards as a compulsive buyer, I suggest that you seek help in dealing with the problems underlying, and created by, this compulsivity. While the scale may overlook those consumers who are not pathological compulsive buyers but whose buying is potentially damaging, it has been used hundreds of times and is considered a reliable indicator of whether you can be classified as a clinical compulsive buyer.

Compulsive Buying Scale

Read each statement and respond—honestly—by circling the number that best represents the extent to which you agree or disagree with each statement. The answers you provide may prove revealing.[20]

1. Please circle the extent to which you agree or disagree with the statement below.

	Strongly agree	Somewhat agree	Neutral	Somewhat strongly disagree	Disagree
a. If I have any money left at the end of the pay period, I just have to spend it.	1	2	3	4	5

2. Please circle how often you have done each of the following things.

	Very often	Often	Some-times	Rarely	Never
a. Felt that others would be horrified if they knew of my spending habits.	1	2	3	4	5
b. Bought things even though I couldn't afford them.	1	2	3	4	5

	Very often	Often	Some- times	Rarely	Never
c. Wrote a check when I knew I didn't have enough money in the bank to cover it.	1	2	3	4	5
d. Bought myself something in order to make myself feel better.	1	2	3	4	5
e. Felt anxious or nervous on days I didn't go shopping.	1	2	3	4	5
f. Made only the minimum payments on my credit cards.	1	2	3	4	5

SCORING INSTRUCTIONS

You will need a calculator to compute your score, but it will be worth it.

1. *Multiply your response to Q1a by .33. Your score:_____*

2. *Multiply your response to Q2a by .34. Your score: _____*

3. *Multiply your response to Q2b by .50. Your score: _____*

4. *Multiply your response to Q2c by .47. Your score: _____*

5. *Multiply your response to Q2d by .33. Your score: _____*

6. *Multiply your response to Q2e by .38. Your score: _____*

7. *Multiply your response to Q2f by .31. Your score: _____*

The total of the above seven scores: _____

Subtract 9.69 from the above total: _____

Write your compulsive buying score here: _____

If your score is less than -1.34, you would be classified as a clinical compulsive buyer. If that's the case, I'm certain you are aware of your problem with spending, but now you need to seek help. If you don't believe the results, ask your spouse, children, or friends if they think you have a problem with spending too much money. If your score was greater than -1.34, you're not off the hook. Remember that compulsive buying is best thought of as a chronic tendency to purchase products in excess of one's needs and resources.

The existence of the treadmill of consumption keeps us buying and buying but getting no closer to happiness than when we began. One logical outcome of all this spending is financial calamity. As we will see in the next chapter, our material longings have left many of us with hardly a financial leg to stand on.

THE CASHLESS SOCIETY

I am having an out-of-money experience.

—AUTHOR UNKNOWN

Are your finances a mess? Are you one of the seven out of ten Americans living paycheck to paycheck? Do you argue with your spouse, children, other family members, or friends about money? Does the term "pay-day loans" ring a bell? Debtors Anonymous (www.debtorsanonymous.org), a nonprofit organization whose primary purpose is to help people free themselves from oppressive debt, has published what they call the twelve "signs of compulsive debting." Please read the list below and place a check alongside any "sign" that applies to you.

1. *Being unclear about your financial situation.* Not knowing account balances, monthly expenses, loan interest rates, fees, fines, or contractual obligations.

2. *Frequently "borrowing" items* such as books, pens, or small amounts of money from friends and others, and failing to return them.

3. *Having poor saving habits.* Not planning for taxes, retirement, or other not-recurring but predictable items, and then feeling surprised

when such obligations come due; having a "live for today, don't worry about tomorrow" attitude.

4. *Falling victim to compulsive shopping.* Being unable to pass up a "good deal"; making impulsive purchases; leaving price tags on clothes so they can be returned; not using items you've purchased.

5. *Having difficulty meeting basic financial or personal obligations* and/ or feeling an inordinate sense of accomplishment when such obligations are met.

6. *Feeling different when buying things on credit than when paying cash.* With credit card use, enjoying a feeling of being in the club, being accepted, being grown up.

7. *Living in chaos and drama around money.* Using one credit card to pay another; bouncing checks; always having a financial crisis to contend with.

8. *Tending to live on the edge.* Living paycheck to paycheck; taking risks with health and car insurance coverage; writing checks and hoping that money will appear to cover them.

9. *Feeling unwarranted inhibition and embarrassment in* what should be a normal discussion of money.

10. *Overworking or underearning.* Working extra hours to earn money to pay creditors; using time inefficiently; taking jobs below your skill and education level.

11. *Being unwilling to care for and value yourself.* Living in self-imposed deprivation; denying your basic needs in order to pay your creditors.

12. *Feeling or hoping that someone will take care of you if necessary,* so that you won't get into serious financial trouble. Expecting that there will always be someone you can turn to.[1]

There are no hard-and-fast rules, but if you checked four or more of the above signs, you need to take a careful look at how you spend and relate to money and then make some serious changes. Actually, if you checked *any* of the twelve signs, it's time to reconsider how you handle money. (Chapters 12 through 14 of this book offer some tried-and-true suggestions to help you get a leg up on your finances.)

As THE CHART BELOW ATTESTS, TODAY'S CONSUMER SOCIETY MAY TRULY have achieved earlier predictions of a cashless society. Simply put, the majority of Americans are broke! According to a 2008 study from National Payroll Week, 70 percent of U.S. families live from paycheck to paycheck, and living hand-to-mouth makes surviving a recession like the current one very difficult. In 2007, the average American put aside 60¢ of every $100 he or she earned, or .6 percent per payday. With the exception of the war years (when there was a lot of investing in U.S. savings bonds), we hit our savings peak in the 1980s at about 10 percent. It's been a downhill slide ever since, with a slight recent upturn (about 4 percent in early 2009) as we witnessed our economy nearly tank. Today's current rate is chilling given that most financial analysts recommend we save 15 to 20 percent of our pretax income for retirement and other goals.[2]

Personal Savings as Percentage of Disposable Income, 1929–2010

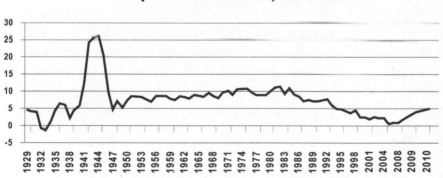

The trilogy of layoffs, the mortgage crisis, and plummeting IRAs—the latter's funds having been withdrawn early, in many cases, because of layoffs and mortgage concerns—has even the most dissolute spenders reconsidering their "financial plans." But not to worry: recessions in the 1970s, in the '80s, and at the start of this new millennium don't seem to have altered our profligate spending.

THE CREDIT CARD CRUNCH

Our profligacy has cost us dearly. Not saving is one thing; piling on debt to finance a spending spree is another. Although many of us would like to blame others, including banks and unscrupulous advertisers, for our current predicament, Columbia University professor David Biems suggests we should look closer to home: "The problem is not the banks, greedy though they may be, overpaid though they may be. The problem is us. . . . We've been living very high on the hog. Our living standard has been rising dramatically in the last 25 years. And we have been borrowing much of the money to make that prosperity happen."[3]

On average, the typical American consumer has a total of thirteen credit obligations on record at a credit bureau. These obligations include an average of about nine credit cards (including store cards and gas cards) and four installment loans (car loans, mortgages, student loans, etc.).[4] The lion's share of that debt is tied to our homes. From 2000 to 2008 mortgage debt rose from $6.9 trillion (that's a thousand billion) to $14.6 trillion in eight short years—an increase of roughly 110 percent. The chart below shows you how much mortgage debt has increased as a percentage of gross domestic product (GDP) for the time period of 1985–2008—in other words, MD/GDP. Note that from 2000 to 2008 the rate of MD/GDP rose from 67 percent to a peak of 106 percent. As financial blogger Bruce Krasting observes:

> The dramatic change in the relationship between mortgage debt and GDP was a significant contributor to the economic expansion during the period. The doubling of mortgage credit fueled the building boom. It was the gasoline that led to the rapid rise in home prices. It is the source of the bubble in both housing

and the broader economy. It is the reason that we are in such a mess today.[5]

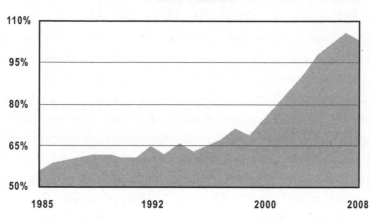

Mortgage Debt as a Percentage of GDP, 1985–2008

The latest data tells us that 20 percent of homeowners owe more on their house than its current value—a situation often referred to as being "underwater" (a particularly apt term) or "upside down." The figure could be closer to 25 percent if house prices drop another 5 percent. Homeownership, long a core element of the American Dream (as we've seen), has now become a nightmare for many Americans. I knew the nightmare was real when many banks and mortgage companies started the practice of using "stated income" when assessing a loan applicant's ability to repay his or her loan. That phrase meant that no attempt would be made to verify an applicant's income—whatever he or she filled in would be considered good enough.

That type of lending practice signified a fundamental shift in how lenders approached their business. Historically, as one might expect, having the borrowers actually repay the loan was crucial to lenders. Believe it or not, the practice of the past ten years or so has strayed from this business model. Until the mortgage crisis (and the government) forced them to stop, many lenders were less interested in whether the loans would be repaid than in the fees and charges they could generate. This emphasis was possible because so much of the consumer debt was being packaged into securities and sold to investors. This was the adult version of the children's game Hot Potato. In the childhood version, you were fine as long as you weren't holding the potato when the time was up. The adult version had an unfortunate twist: it wasn't just the people holding the potato who got burned, but everyone who played—and even those who didn't! The entire country is currently suffering because of such nonsense. Practices that produced record profits for most of these lenders nearly led to the collapse of our nation's financial system. This perversion of the American Dream—the idea that everyone should own a home, regardless of his or her ability to pay—nearly spelled an end to the American Dream itself.

Mortgages aren't our only financial obligations, of course. Total U.S. consumer debt, which does not include mortgage debt, reached $2.56 trillion by the end of 2008. This is a 22 percent increase since 2000. Household debt, including *both* mortgages and credit card debt, according to the Federal Reserve Board now represents 19 percent of household assets, compared with 13 percent in 1980. Over the past twelve months, 26 percent of Americans (58 million people) reported that they were not able to pay all of their bills on time. This number rose to 51 percent, almost double, among African Americans.[6]

Easily available credit has been our Achilles' heel, as a nation. I remember the quaint old days when we used cash to pay for a fast-food meal, a gas fill-up, our dry-cleaning, and—heaven forbid—even our weekly or monthly groceries. Now all of this and more can be paid for with plastic. Credit card marketers have focused much of their recent attention on convincing consumers to buy even the smallest of items with their credit cards, resulting in "micro-payments." Got a speeding ticket? No problem: many state troopers carry card swipes in their patrol cars. Vending machines, subway systems, and even charities accept plastic. The federal government has gotten involved too, handing out cards in lieu of food stamps and child-support checks. Hip-hop mogul and entrepreneur Russell Simmons (and others) provides a service—for a fee, of course—that puts people's paychecks directly onto a credit card, allowing even disadvantaged souls who don't have bank accounts access to the wonder of credit cards.

Let me throw out some credit statistics here—the results of a great many studies by various researchers (each measuring a different factor or spanning a different timeframe). At last count, there were about 1.5 billion credit cards in use in the United States. If stacked one on top of another, all of these credit cards would extend beyond seventy miles into outer-space—about the height of thirteen Mount Everests. More specifically, by 2008, there were 339 million Visa credit cards and 314 million Visa debit cards, 263 million MasterCard credit cards and 126 million MasterCard debit cards, 59 million American Express cards, and 57 million Discover cards. According to the U.S. Census Bureau, there were 159 million credit card holders in the United States in 2000 and 173 million in 2006, and that number is expected to swell to 181 million users by 2010. Overall, U.S. consumers held an average of 5.4 cards in 2008.[7]

By the end of 2008, Americans' credit card debt had reached $972.73 billion. For the 75 percent of households that owned a credit card, that translated to an average outstanding credit card debt of $10,679 in 2008. Approximately 60 percent of these households carried a balance from month to month.[8]

In 2007, the average balance for those households that carried a balance rose 30.4 percent. Twenty-eight percent of all households say their ability to pay off their credit card balance has become more difficult. In 2008, 15 percent of American adults, nearly 39 million people, were late making a credit card payment. From 1989 to 2004, the percentage of card holders incurring late fees due to late payments of sixty days or more jumped from 4.8 percent to 8 percent.[9]

Nearly as devastating are the one in six American families who pay only the minimum payment due each month. In the 1970s, the average minimum payment required was about 5 percent; by 2000 it was around 2 percent. Lower minimum payments, along with reward programs and enhanced consumer services, is a practice used by credit card issuers purportedly to encourage customer loyalty. Seen from a more cynical perspective, reducing the minimum payment is an effective technique to increase card-issuer profitability by preserving outstanding balances. From 2000 to 2008, consumer "revolving debt" (that is, debt that isn't paid off each month) increased by over $281 billion, reaching $960 billion—an approximately 40 percent increase.[10]

The Bankruptcy Abuse Prevention and Consumer Protection Act of 2005 required new disclosures on the front of each periodic statement received by credit card holders warning that making only the minimum payment increases both the total interest paid and the time it takes to repay the loan. A hypothetical example must follow the warning, along with a toll-free number through which card holders can get an estimate of the time it would take to pay off their account balance.

I think most of us know, but choose to ignore, the financial impact of making only minimum payments on our credit cards. Just for grins, here's a simple example of the difference that results from increasing the minimum payment from 2 percent to 4 percent. A card holder with a $10,000 balance in a payment program taking advantage of a 2 percent minimum and a 16 percent interest rate would need more than forty years to pay off the balance and would pay a total of $19,329 in interest. The same card holder making a 4 percent minimum payment would pay off his or her balance in approximately fourteen years and would pay a total of $4,931 in interest. Not a good deal, but certainly better than the 2 percent minimum scenario.

DOES MONEY BURN A HOLE IN YOUR POCKET?

As bleak a picture as the above statistics might paint, there are still those among us who have avoided the credit trap, have manageable mortgages, drive paid-off cars, and have a little socked away for our kid's college and our own retirement. It's time for a little self-reflection: Do *you* fall into this latter category or into the bleak earlier picture?

How we approach the spending of money plays an important role in our ability to manage our financial obligations and reserves. Please take a couple of minutes to answer the three questions below, which are part of what the authors, Scott Rick, Cynthia Cryder, and George Lowenstein, call the "Spendthrift-Tightwad Scale."[11]

Spendthrift-Tightwad Scale

1. Which of the following descriptions fits you best?

1	2	3	4	5	6	7	8	9	10	11

Tightwad (difficulty spending money)		About the same or neither		Spendthrift (difficulty controlling money)

2. Some people have trouble limiting their spending: they often spend money—for example, on clothes, meals, vacations, phone calls—when they would do better not to.

    Other people have trouble spending money. Perhaps because spending money makes them anxious, they often don't spend money on things they should spend it on.

    a. How well does the first description fit you? That is, do you have trouble limiting your spending?

1	2	3	4	5
Never	Rarely	Sometimes	Often	Always

    b. How well does the second description fit you? That is, do you have trouble spending money?

1	2	3	4	5
Never	Rarely	Sometimes	Often	Always

3. Following are two scenarios describing the behavior of two shoppers. After reading about each shopper, please answer the question that follows.

Mr. A is accompanying a good friend who is on a shopping spree at a local mall. When they enter a large department store, Mr. A sees that the store has a "one-day-only sale" where everything is priced 10 to 60 percent off. He realizes he doesn't need anything, yet he can't resist and ends up spending almost $100 on stuff.

Mr. B is accompanying a good friend who is on a shopping spree at a local mall. When they enter a large department store, Mr. B sees the store has a "one-day-only sale" where everything is priced 10 to 60 percent off. He figures he can get great deals on many items that he needs, yet the thought of spending the money keeps him from buying the stuff.

In terms of your own behavior, who are you more similar to, Mr. A or Mr. B?

1	2	3	4	5
Mr. A		Neither		Mr. B

This scale's authors argue that consumers rely on an immediate "pain of paying" to control their spending. You know the types at both ends of the pain continuum: spendthrift and tightwad. Think of the uncle whose money always seems to "burn a hole in his pocket"—here today, gone tomorrow. Spending money is painless for the spendthrift. On the other end of the continuum, remember the uncle, parent, brother, or sister who needed a crowbar to part with money. *That* person is at the tightwad end of the continuum. It just hurts too much to part with money, even when its expenditure is well justified. In the middle, we have the "unconflicted." These consumers experience intermediate levels of pain compared to spendthrifts and tightwads—just the right level of pain to carefully consider spending decisions and make decisions that are consistent with their long-term best interests. Not a profligate lifestyle, but not a miserly one either.

So how did you do? Use the following norms to peg where you fall along the spendthrift-tightwad (ST-TW) continuum. The average score on the ST-TW scale is 14.38. Note: You must reverse your scores for items 2b and 3 (for example, 5=1, 4=2, 2=4, 1=5).

Does your score accurately reflect your "pain of paying" when making spending decisions? Does money flow through your fingers like water off a duck's back, or is it supremely painful for you to spend money even when you really need something? Or, are you the rare bird who carefully weighs most spending decisions and usually ends up making good decisions?

SPENDTHRIFT-TIGHTWAD SCALE NORMS

Tightwads = 4 to 11
Unconflicteds = 12 to 18
Spendthrifts = 19 to 26

Using the ST-TW scale, Rick, Cryder, and Lowenstein found that tightwads and unconflicteds are better savers. Not surprisingly, spendthrifts carry more credit card debt than tightwads and unconflicteds. Tightwads are not necessarily happier, however. Their high levels of pain of paying often lead to a miserly existence and high levels of anxiety and stress when dealing with money. As with many things in life, moderation may be the key. It may be that the relationship between ST-TW scores and happiness is curvilinear, as depicted below—in other words, those who experience a moderate level of "pain of paying" may be the happiest.

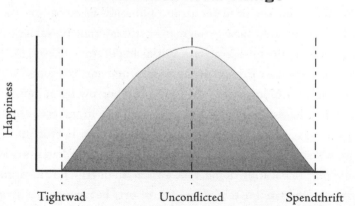

Moderation in All Things

It should come as no surprise that materialistic people have a different attitude toward and relationship with money than less materialistic people. Research has found that those high in materialism typically feel that they need an income about 50 percent higher than do those low in materialism. As discussed earlier, materialists set an unrealistically high standard of living and spend in an attempt to achieve that standard. Materialists consume more than they can afford because material possessions play a central role in their lives, because those possessions are both seen and projected as indicators of success, and, most importantly, because materialists believe that consuming will make them happy. In their attempt to be happy through consumption, highly materialistic people spend more and save less money than less materialistic people.

University of Canterbury (New Zealand) researcher John Watson studied the relationship between materialism, spending tendencies, saving, and debt in a large sample of adults living in Pennsylvania—a sample that mirrored reasonably well the typical U.S. consumer.[12] In his mail survey, Watson collected data on measures of materialism, propensities to spend/save, attitudes toward debt, and behavioral measures of spending and saving (for example, did the respondents have savings or checking accounts, investments, and installment credit—that is, credit cards and loans). Respondents whose materialism score on the eighteen-item materialism scale placed them in the top 25 percent of the sample were considered to be highly materialistic, and those scoring in the bottom 25 percent were deemed to be low in materialism. This split ensured that the two groups being compared varied widely in their materialistic values.

The results confirmed Watson's initial hypotheses. Respondents in the highly materialistic group saw themselves more as "spenders" and had more favorable attitudes toward the long-term borrowing of money for a myriad of purchases than did those who rated low in materialism. They reported, for example, that they would be willing to borrow money to purchase home furnishings, have optional or corrective dental work, buy a second home or vacation home, take a vacation or trip, buy a swimming pool, take a honeymoon trip, stay at a resort, buy expensive sporting goods, buy furs or jewelry, invest in art or other collectibles, or buy a recreational vehicle, camper, or boat. Clearly, materialists have favorable attitudes about borrowing money, especially for luxury items.

Recent research by Julie Fitzmaurice of Merrimack College confirmed the highly materialistic consumer's penchant for luxury items.[13] In studying the relationship between "splurge" purchases and materialism, Fitzmaurice found that materialists were more likely to splurge on items they wore and items they referred to as expensive. A "splurge" was defined as a purchase that is (1) desired yet not necessary, (2) self-indulgent, (3) outside the normal realm of purchase, and (4) considered loose spending. Highly materialistic people considered visible possessions as the most important. Fifty-seven percent of splurge purchases by the highly materialistic respondents were made to enhance the purchaser's appearance, as opposed to 20 percent for those low in materialism. Interestingly, materialists also felt more guilty and irresponsible when reflecting on their splurge purchases.

This willingness to borrow in the highly materialistic group has obvious behavioral consequences. Fitzmaurice found that these people carried higher balances on their credit cards than their less materialistic counterparts and were more likely to carry over balances, own more credit cards, and pay finance charges on these same cards. High-materialism respondents were also more likely to have installment loans. As this research shows, the perceived urgency of numerous wants for consumer goods and experiences stacks the deck against saving and in favor of spending for the highly materialistic consumer.

SPENDING OUR WAY TO UNHAPPINESS

The overarching objective of this chapter is to explain how materialism, through its impact on how we manage our financial affairs, affects our life satisfaction (or "subjective well-being" or "happiness"). A number of quality-of-life studies have shown that a person's life satisfaction can be explained and predicted by how satisfied people are with certain important life domains—that is, important areas in our lives that affect our well-being. The hierarchy of life domains below is a simple model that helps explain life satisfaction. Its basic premise is that overall life satisfaction (at the top of the pyramid) is predicated upon one's satisfaction with each of life's domains: the greater your satisfaction with each of the life domains, the greater your overall life satisfaction.[14]

Hierarchy of Life Domains

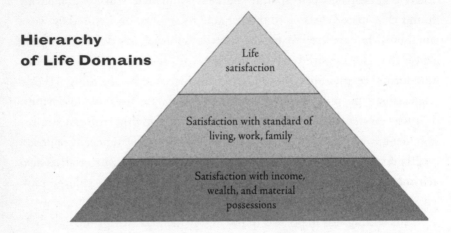

This approach is called "bottom-up vertical spillover" by researchers in the field. Given the current economic recession and poor money management skills of people high in materialism, vertical spillover is a reality for many of us. As stocks plunge and companies go under, reports the *Washington Post*, 88 percent of Americans are worried about the economy's health and direction.[15] Less than half of U.S. adults now believe that they have sufficient funds heading into retirement; about two-thirds worry about their own family's economic situation. Clearly, financial matters both under our control (our spending habits) and outside of our control (the economy) play an important role in our overall life satisfaction.

There is also a process called "top-down vertical spillover." This occurs when our overall life satisfaction affects how we evaluate, for example, our standard of living. If we are satisfied with our life in general, we may be more likely to positively evaluate our standard of living or other life domains.

Finally, "horizontal spillover" is the label given when spillover occurs between life domains at the same level in the hierarchy of life domains depicted above. Our life satisfaction may be influenced by the impact of one life domain—let's say money—on other life domains, such as wealth or material possessions. If we're never happy with how much money we earn, this lack of satisfaction can negatively impact how satisfied we are with our standard of living or our material possessions. Horizontal spillover may be particularly relevant for materialists, who (as noted earlier) are chronically dissatisfied with their level of income and its ability to secure them the possessions they desire. They consistently report a higher level of needed income compared to less materialistic people. This dissatisfaction can have both vertical and horizontal spillover effects. Unhappy with their income, materialists are less satisfied with their overall quality of life (vertical spillover). Additionally, dissatisfaction with income generally impacts satisfaction with material possessions and wealth (horizontal spillover).

A RECENT STUDY BY THE TRIO OF JING JIAN XIAO OF THE UNIVERSITY OF Rhode Island and Chuanyi Tang and Soyeon Shim of the University of Arizona investigated the impact of positive financial behaviors on financial satisfaction and financial satisfaction's impact on life satisfaction.[16] In essence, the researchers were testing whether satisfaction with one's finances spilled over (vertically) into one's overall life satisfaction, as earlier research in England, the United States, and Canada had suggested.

The researchers measured positive financial behaviors by looking at ten practices related to cash, credit, and saving management. Relevant to the present discussion, such behaviors included "saving money regularly," "setting aside money for emergencies," "contributing to an investment or retirement account," and "spending with a budget." The study's participants were also asked about their level of satisfaction with their finances and responded to a commonly used five-item scale that measures overall life satisfaction.

A Web-based survey of 976 college students was used by these researchers to test the following hypotheses:

1. Performing positive financial behaviors would be positively associated with financial satisfaction.

2. Higher debt level would be negatively associated with financial satisfaction.

3. Higher financial satisfaction would be positively associated with life satisfaction.

As hypothesized, positive financial behaviors such as paying off credit cards each month, paying bills on time, and maintaining a sufficient balance in a bank account, as well as saving behaviors such as setting aside money for emergencies and saving money regularly, were positively associated with financial satisfaction. In turn, higher levels of financial satisfaction led to an increase in overall life satisfaction. Student debt, however, was not related to financial satisfaction when positive financial behaviors were performed. It appears that with debt as a given for financing an increasingly expensive college education, it has little bearing on a student's financial satisfaction.

PAPER OR PLASTIC?

An interesting additional finding of Rick, Cryder, and Lowenstein's research on spendthrifts and tightwads is that more people were classified as tightwads than spendthrifts. With the typical American's financial affairs in such disarray, how can this be? It all comes down to how we pay for our purchases and to the "pain of paying" associated with our payment method of choice. We are a nation addicted to plastic, as earlier statistics showed. Using credit cards to pay for our purchases greatly reduces the pain of paying—so much so, in fact, that when credit cards are used, the spending habits of spendthrifts and tightwads converge. Even tightwads loosen up the old purse strings when using plastic to pay for their purchases. The balance of this chapter will discuss why credit cards have earned the name "spending facilitators" by those of us who do research in this area.

We have a lot of payment options available to us when paying for our purchases. There's always cold, hard cash (pretty painful), as well as debit cards, checks (quaint, but a bit antiquated), and the ubiquitous credit cards. When choosing which option to use, few of us stop to think that how we pay for purchases might affect our purchase decisions. Yet research has shown that we spend significantly more when we use credit cards. Furthermore, when we use credit cards we make quicker purchase decisions, are more likely to buy something, and are more likely to be willing to pay a higher price for it. Even at fast-food restaurants, when we use credit cards instead of cash we spend anywhere from 60 to 100 percent more. As I mentioned in chapter 1, this "supersize effect" of credit cards caused the average bill at McDonald's to increase from $4.50 to $7.00 when customers used credit cards instead of cash.[17]

THE COST OF CREDIT   Americans' love of credit cards is costing us a lot of money—to the tune of about $50 billion annually. As you may know, retailers pay transaction fees to credit card companies when you use your Visa, MasterCard, or other credit cards to pay for your purchases. This extra cost gets passed directly back to you in the form of higher prices. For example, a 3.5 percent credit card fee on an $800 flat-screen TV is $28.08. A $7 fast-food sandwich racks up fees of about $0.36. (There's a minimum per-item fee on most purchases.) A typical sandwich shop— a Subway franchise, for example—will pay about $10,000 each year in credit card processing fees. That's a whole lot of baloney![18]

!! CONSUMERS !!

!! GONE WILD !!

In 1986, the *Journal of Consumer Research* published a paper by Richard Feinberg that investigated the impact credit cards have on spending decisions.[19] Even by the mid-1980s it was commonly accepted by retailers, credit researchers, writers, and consumers themselves that people spent more when they used credit cards. Feinberg set out to see why. Operating under the assumption that credit card stimuli (card logos, for example) are so closely associated with spending that they increase the probability of spending, increase the amount spent, and reduce the time it takes to decide to spend, Feinberg conducted several experiments to test his hypotheses.

The rationale behind Feinberg's series of experiments was fascinating. Earlier research on what one would assume is a totally different subject had found that the mere presence of weapons triggered more aggressive responses in study subjects than were seen when no weapons were present. Feinberg thought that this finding, labeled the "weapons effect," was analogous to the impact that credit card stimuli would have on spending.

Tip Size: Cash vs. Credit Card

	SIZE OF CHECK			
	1	2	3	4
	$17.00	$23.00	$31.00	$47.00
Credit card	$3.03	$3.92	$4.91	$8.04
Tip Size (in percentages)	(17.8%)	(17.1%)	(15.8%)	(17.1%)
Cash	$2.36	$3.60	$4.64	$7.23
Tip Size (in percentages)	(13.9%)	(15.6%)	(14.9%)	(15.4%)

In an initial study, Feinberg investigated whether tips left by customers would be different depending on whether they were made by cash or credit card. One hundred and thirty-five customers of a local restaurant were observed at random times over a one-week period. The size of the check, number of people in each party, method of payment (cash or credit card), and amount of tip were carefully recorded by the waitstaff. It was anticipated that credit card stimuli present at time of payment would increase the size of the tip. The table above provides support for such thinking. At each level of bill size, those who paid with credit cards left bigger tips. Overall, tips left on a credit card averaged 16.95 percent compared to 14.95 percent when left in cash.

To verify that it was the use of credit cards that facilitated bigger tips, a series of four additional experiments were conducted. In the first experiment, sixty subjects (college students) were assigned to one of two groups—(1) credit card stimuli present or (2) credit card stimuli absent. Subjects were asked to look at a loose-leaf booklet titled "Consumer Products" that contained pictures of seven products, and they were instructed that they would later be asked for information about those items. The pictures of the products were pasted in the center of plain white pages and covered in clear plastic. The products were clearly labeled at the top of the page as product 1, product 2, etc.

For half of the randomly assigned subjects, a MasterCard logo was visible on the upper left-hand corner of the table near the book. These subjects were informed ("cued," in research vernacular) that the credit card logo was from a previous experiment. The other thirty subjects, shown the identical booklet, saw no credit card logo. Subjects were questioned about how much they would be willing to spend for each item, and then

Willingness to Pay

PRODUCTS	CREDIT CARD LOGO PRESENT	LOGO NOT PRESENT
Dress #1	$41.50	$27.77
Dress #2	33.91	21.09
Tent	77.73	69.96
Man's Sweater	20.64	13.91
Lamp	40.41	28.36
Electric Typewriter	165.36	131.45
Chess Set	43.14	35.29

(to hide the true purpose of the experiment) they were also asked to write down the most distinctive aspect(s) of each product. As can be seen from the table above, subjects who were exposed to the credit card logo consistently said they would be willing to spend more for each of the products, compared to subjects when no credit card logo was present.

In a second experiment, twenty-four female undergrads were assigned to one of two experimental conditions—credit card logo present or absent. Each subject was led to a table in a room. A slide projector was placed on the table immediately in front of them, along with a button labeled "response." In a format similar to that of the previous experiment, subjects were instructed that they would be evaluating a number of consumer products that would be projected on the screen in front of them. They were told that they could look at the slide for as long as they wanted and were to push the "response" button when they had decided how much they would be willing to spend for each product; they were then to write this amount on an answer sheet provided for them.

Subjects were shown twelve consumer products, which included such items as toasters, TVs, digital clocks, and stereos. A timing clock started when each slide popped up and stopped automatically once the subject pressed the "response" button. For one subgroup of respondents, a MasterCard logo was present in the upper-left-hand corner of each slide. Subjects who were exposed to that logo reported that they were willing to spend more for all of the products. Additionally, the presence of a credit card logo reduced the decision time for making a purchase. In both these experiments, then, the presence of a MasterCard logo increased the magnitude of estimated spending, and in the second one (which measured response time) it reduced the time needed to make such decisions.

In a third experiment, forty subjects were randomly assigned to perform the experiment either in the presence or absence of a credit card logo. Subjects were led to an office under the cover story that they would be participating in a study about "impression formation." Subjects were given a short description of a person and asked to form impressions based upon this information. The true intent of the experiment, however, was to see whether the presence of a credit card logo would affect the participants' willingness to make a charitable donation. For half of the subjects, a credit card logo was visible on the upper-left-hand corner of the table where the students formulated their "impressions." Ten minutes after each subject arrived at the office, a stranger knocked on the door, approached the subject, and explained that the United Way charity was conducting a door-to-door survey to assess the viability of soliciting donations on campus. The solicitor then asked how much the subject would be willing to donate to United Way, if asked. The study's hypothesis was supported: subjects where a credit card logo was present reported being willing to donate on average $4.01 compared to $1.66 for subjects where no credit card logo was present.

In a fourth experiment, thirty undergraduates were assigned to one of two individual testing sessions—one where a credit card logo was present and another where it was not. This time, the MasterCard logo and regular- and large-size replicas of actual MasterCards were placed on the upper-right-hand corner of the subjects' table. As in the previous experiment, subjects were led to an office under the guise of an "impression formation" exercise. This time, however, the United Way solicitor asked for an actual donation. As hypothesized, those subjects where a credit card logo was present made larger donations than those where no logo was present. Additionally, the time needed to make the decision to donate money was considerably shorter when a credit card logo was present.

Feinberg's experiments provide substantial support for the simple premise that people spend more money with credit cards. The mere presence of credit cards leads us to be more likely to spend, more willing to spend more, and faster at making our spending decisions. The more we use credit cards to make purchases, the more we become "conditioned" to spend, so that eventually credit card logos and credit cards themselves gain the ability to elicit spending on their own—a frightening scenario that portrays consumers as unthinking zombies conditioned to spend mechanically when exposed to credit card stimuli.

The above results are consistent with earlier findings of Feinberg and others, who confirmed that mode of payment impacts our spending. Any payment method that decreases the "pain of paying" (not just credit cards but using gift cards, PayPal, or even debit cards) can lead to overspending. So why is that? Let's look at the latest research that attempts to explain why we spend more when we use credit cards.

Research by marketing professor Dilip Soman that was reported in the 2001 volume of the *Journal of Consumer Research* does an excellent job of explaining the facilitating impact of credit cards on spending.[20] Soman argued (and produced evidence to support that argument) that the form of payment mechanism impacts the recollection of past payments and, therefore, influences future expenditures. Simply put, compared to payment by cash, checks, and even debit cards, credit card payment is, in Soman's words, less "salient" and "vivid" (memorable). The "buy now, pay later" nature of credit cards makes it less painful ("salient") to spend money when making purchases with them, compared to with cash, checks, and debit cards, because the impact of these latter three on one's financial position is immediate; paying in cash is downright painful and immediately leaves purchasers with less money in their pocket. The delayed impact on one's wealth when paying with a credit card leads to a lower aversive impact; the delay built into the billing cycle separates the enjoyment of the purchase from payment, which leads to a softening of the payment's impact. The impact is lessened not only by that delay, but also by the fact that large check and debit card charges in other accounts may be more "salient" and by the lumping of a particular credit card payment among a multitude of other credit charges.

Credit card payments are also less memorable because of the lack of what we might call "rehearsal." When counting out cash or writing a check, the consumer has an opportunity to learn and remember the final price paid. In writing a check, consumers write down the total twice—in words and figures. The same rehearsal holds for cash: you have to count the cash (a very painful task) when making payment and again when your change is returned. This repetition makes it much more likely that you'll remember how much you spent. With credit cards, however, you only have to sign a receipt with a printed total already provided. What's more, in many instances when the total is under $25, you aren't even asked to sign a receipt. Many of us don't even ask for a receipt when paying for gas or other purchases; we don't worry about it until the credit card bill arrives. It's this combination of the lack of rehearsal and the delayed impact on our wealth (resulting together in a less aversive or painful experience) that leads us to spend more when we use credit cards. We overestimate the amount of available wealth, which in turn increases the likelihood of making additional purchases.

Soman conducted two real-world studies to test his hypothesis that credit cards negatively impact our ability to recall how much we spend as consumers. The first study is my favorite. Forty-one students were intercepted immediately after purchasing books and other supplies at the campus bookstore. The students were asked how they had paid for their purchases and to recall the exact amount they had spent. They were then asked to confirm the amount they had spent by checking their receipts. Of the eighteen students who had paid with cash, twelve (or 66.7 percent) accurately recalled the amount they had spent, and the remaining six were within $3 of the true amount. It wasn't as pretty for those students who paid by credit card: of the twenty-three who paid by credit card, only eight (or 34.8 percent) could accurately recall the amount; the remaining fifteen either reported an amount lower than the true amount or confessed that they had no idea how much they had spent.

In a second study, Soman asked a sample of thirty single-income-family individuals with only one credit card to bring the experimenters their unopened credit card bill as soon as it arrived. They were also instructed to save and bring in receipts for all significant purchases (over $20) during the same period, regardless of whether they were paid for by cash, check, or credit card.

When the participants showed up, they were first asked to recall all of their expenses and then to open their statement and receipts and write down the itemized expenses. Without exception, all thirty participants underestimated their credit card expenses to a greater degree than their cash and check expenses. In total, respondents underestimated their credit card expenses by an average of 25 percent, compared to 7 percent for purchases made by cash or check. Both studies provide evidence that credit card payments are relatively less salient and memorable than payments made by check or cash.

Credit card use among college students has become a topic of growing concern over the past decade. Sallie Mae, the college loan behemoth, recently released its 2009 report, entitled "How Undergraduate Students Use Credit Cards." This is the fifth credit card usage study in a series that began in 1998. The sample consisted of 1,200 randomly selected student loan applicants between the ages of eighteen and twenty-four. The unsettling results of the Sallie Mae study document an increasing use of credit cards among the college student population. The table below highlights some of the most telling findings of the study.

Credit Card Trends Among College Students, 1998–2008

	1998	2000	2002	2004	2008
Percentage who have credit cards	67%	78%	83%	76%	84%
Average number of credit cards	3.50	3.00	4.25	4.09	4.60
Percentage who have four or more cards	27%	32%	47%	43%	50%
Average credit card debt	$1,879	$2,748	$2,327	$2,169	$3,173
Median credit card debt	$1,222	$1,236	$1,770	$946	$1,645
Percentage with balances of $3,000–$7,000	14%	13%	21%	16%	21%

Source: 2009 Sallie Mae report: "How Undergraduate Students Use Credit Cards," www .salliemae.com.

Additional tidbits regarding college students and credit cards, assessed in 2008 (and taken from that same source), include the following:

- Among freshmen arriving on campus, 39 percent already had a credit card—up from 23 percent in 2004.

- Only 15 percent of freshmen had a zero balance, compared to 69 percent in 2004.

- Seniors were graduating with $4,100 in credit card debt; about one-fifth of seniors carried balances over $7,000.

- Nearly one-third of all students paid for tuition with credit cards.

- A whopping 60 percent of students experienced surprise at how large their credit card balance had grown, and 40 percent charged items knowing that they didn't have the money to pay for them.

- Only 17 percent of the students regularly paid off their credit card balance each month.

- Reassuringly, 84 percent of students indicated that they needed more education on financial management topics.

The authors of the report conclude that "many college students seem to use credit cards to live beyond their means."[21] All this de facto borrowing has taken a psychological toll on students. The same study found that 24 percent of students reported that they were "extremely anxious" (level 1), and another 21 percent reported that they were "very anxious" (level 2) about their ability to pay their monthly credit card bills. College students who use credit cards irresponsibly suffer both financially and psychologically. Students with higher consumer debt earn poorer grades, drop out of school at a higher rate, suffer more from depression, file more frequently for bankruptcy, and work more hours to pay their bills. Credit card debt has also been linked to a number of suicides by college students.

One college administrator was quoted as saying, "This is a terrible thing. We lose more students to credit card debt than academic failure."[22]

Whether we're young people just starting out in college or mature adults, our obsession with credit cards is a natural outgrowth of our increasing materialism. Credit cards allow us to shop impulsively, indulge our every whim, and postpone worrying about the consequences until another day. As this chapter attests, paying by credit card or using any other payment method that reduces the pain of paying has a devastating impact on our ability to manage our finances; indeed, taking advantage of the ready availability of credit cards is akin to throwing gasoline onto a fire.

We've seen that profligate spending runs counter to our financial well-being. Now chapter 7 will extend the negative impact of materialism beyond the financial realm and explain how materialistic life goals lead to less satisfying lives.

MONEY'S HIDDEN COSTS: SACRIFICING OUR LIFE GOALS

If you don't know where you're going, any road will get you there.

—AUTHOR UNKNOWN

There has been a dramatic shift in the goals and values of the American people over the past half-century. It's not something that happened overnight or even over the course of a few years; instead, it has crept up on us slowly and quietly, like a thief in the night.

The past fifty years have seen a steady upward trend in materialism. The clearest evidence of that trend can be found in an annual survey of college freshmen conducted by UCLA and the American Council on Education.[1] Each year approximately 700 two-year and four-year colleges and universities administer the survey to over 400,000 freshly scrubbed entering freshmen during orientation or registration. The survey covers a broad array of topics, but of current interest to us are the twenty "life goals" that respondents are asked to rate, using a scale that ranges from "essential" to "very important" to "somewhat important" to "not important."

The life goals addressed in the survey include things like "becoming an authority in my field," "raising a family," "being very well-off financially," "helping others who are in need," "adopting green practices to protect the environment," and "developing a meaningful philosophy of life." Interestingly, "being very well-off financially" outranked all of the other life goals in 2010, with 77 percent of freshmen responding that it was either essential or very important to them. As can be seen from the table below, this was not always the case. In 1975, approximately 60 percent of freshmen felt that being well-off financially was essential or very important. The trend line, however, has shown a consistent increase in materialistic values among our best and brightest. In contrast, the life goal of "developing a meaningful philosophy of life" has consistently lagged behind more pecuniary pursuits.

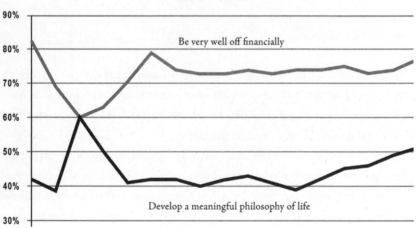

Life-Goal Trends Among College Students, 1965–2010

Skepticism about the ability of money and material possessions to bring about lasting happiness can be found in the tenets of humanistic psychology (HP). The luminaries of HP include Carl Rogers, Abraham

Maslow, Eric Fromm, and Rollo May, to name just a few. At the very core of HP is the struggle of humans to grow and develop. Although proponents of HP agree that some basic level of food, shelter, and clothing is needed, they also feel that the effort put forth to constantly improve upon these needs can stand in the way of psychological health.[2]

Factors such as self-acceptance, the ability to form and maintain intimate relationships, and involvement in the community are held out by HP proponents as the cornerstones of psychological well-being. A highly materialistic lifestyle, contend many Humanistic psychologists, is a double whammy. Materialists are distracted from experiences that may make them happy and also miss out on the opportunity to live a more meaningful life.

A good example of the negative impact of materialism can be seen in two-wage-earner families. It's common today for both spouses to spend a majority of their time working so that they can purchase what the consumer culture has on offer. No amount of shiny objects, however, can make up for missed time with your children, spouse or friends, or fun activities. It is such missed opportunities that undermine the happiness of materialistic people.[3]

From an HP perspective, self-actualizing humans are the rare breed who have the ability to assess their underlying condition and choose a course of action that enables them to adapt and continue to develop. Think back to Psychology 101 and Abraham Maslow's "hierarchy of needs."[4] According to that hierarchy, the ultimate goal is self-actualization—the point at which a person wants (and is motivated) to fulfill his or her maximum potential. Other lower-level needs must be met, however, before a person can strive for self-actualization. People who reach that peak of the hierarchy find themselves in a place where they are driven by an inner voice rather than by lower-level survival needs or the dictates of others. Visualize yourself climbing the levels of Maslow's hierarchy. Where are you along this journey? Maslow himself acknowledged that the drive for self-actualization is a relatively faint voice competing to be heard among the din of lower-level needs.

Maslow's Hierarchy of Needs

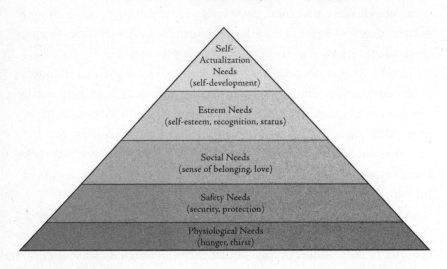

Self-
Actualization
Needs
(self-development)

Esteem Needs
(self-esteem, recognition, status)

Social Needs
(sense of belonging, love)

Safety Needs
(security, protection)

Physiological Needs
(hunger, thirst)

Jean-Paul Sartre, in his book *Being and Nothingness*, creates a different (but similar) hierarchy, making a distinction among three levels of existence: having, doing, and being. In *having*, the lowest of the three levels of existence, people are preoccupied with building wealth, with the accumulation of material possessions. In *doing*, they are preoccupied with lifestyles and activities rather than with material possessions and money. In *being*, Sartre's highest level of existence, people find fulfillment and serenity in who they are rather than in what they have or do.

In today's consumer culture, money and the pursuit and accumulation of material possessions are keeping us from moving upward on Maslow's hierarchy or on Sartre's levels of existence. In other words, materialistic pursuits are preventing us from achieving the things in life that can deliver lasting happiness. It is my hope that this current discussion of life goals will support movement in that direction. The critical question at this juncture is this: Do we, as individuals, have goals and values that further our well-being? I'll give you a little background to help you answer that question for yourself.

Research has identified two broad classes of goals: extrinsic and intrinsic.[5] Extrinsic goals include financial success, social recognition, and physical attractiveness. Ultimately, meeting these goals is contingent upon the approval of others, and their pursuit often results in anxious and unhappy people. Extrinsic goals distract people from the important underlying psychological needs discussed above—needs for actualization and fulfillment. Additionally, extrinsic goals lead people to engage in activities that are less enjoyable and often times stressful and anxiety-provoking. For example, pursuing a goal of financial success entails working long hours, often under stressful circumstances, in the knowledge that even with the requisite effort the goal may not be achievable. And don't forget about the so-called opportunity costs of such a goal—time away from loved ones and inadequate time for self-development and community involvement. The mass pursuit of extrinsic goals is a large part of why consumption in the United States has grown over the past forty years with no attendant increase in our happiness.

GEN NEXTERS Generation Next consists of people between the ages of eighteen and twenty-five years old in 2006 (born between 1981 and 1988). The Pew Research Center interviewed 579 people in this age range as part of a larger survey of 1,501 U.S. adults. The Gen Nexters were asked about the most important goals for people in their age group. A large portion, 81 percent, responded that being rich was either the most or second-most important goal for their generation. Second on the list of life goals for Gen Nexters was being famous. Among this subset of respondents, 51 percent said that being famous was either their most important or second-most important goal. Much further down the list, 12 percent of Gen Nexters felt that helping people in need was their generation's

!! CONSUMERS !!

!! GONE WILD !!

most important goal, followed by being leaders in their community (17 percent) and becoming more spiritual (4 percent). I hope that this focus on wealth and fame is just a phase.[6]

Intrinsic goals, on the other hand, include self-acceptance, self-esteem, psychological growth, affiliations (in other words, relationships with family and good friends), and community involvement. These goals focus on satisfying the deep-seated needs we share as humans. These include the need to feel competent, self-sufficient, to have healthy relationships, to expand one's horizons, and to be part of something bigger. People with intrinsic goals are more likely to choose activities and lifestyles that encourage personal growth and healthy relationships as well as increased community involvement. It is just such activities that contribute to one's personal growth and an overall heightened sense of well-being.

Pioneering work by researchers Tim Kasser and Richard Ryan investigated whether the content of a person's goals impacted his or her well-being.[7] In the first of two studies conducted in the 1990s, Kasser and Ryan developed what they called an "Aspiration Index." Initially, the index included twenty-one statements that measured how important each of the aspirations ("goals") was to respondents and how likely they felt it was that they would attain each one in the future. Three intrinsic goals and one extrinsic goal were measured in the first study. The three intrinsic goals measured included self-acceptance (four statements), affiliation (six statements), and community feeling (six statements). Relevant to our current discussion, the one extrinsic domain of aspirations was labeled financial success; that domain contained five statements measuring how important financial success was to the participants and how likely they felt they would be able to attain each financial success–related aspiration. In 1996 the pair added two additional extrinsic goals: social recognition and attractive appearance. The statements used to measure each of the six aspirations—the original four plus the added two—are listed below.

Aspiration Index

INTRINSIC GOALS

Self-Acceptance

- You will be the one in charge of your life.

- At the end of your life you will look back on your life as meaningful and complete.

- You will deal effectively with problems that come up in your life.

- You will know and accept who you really are.

Affiliation

- You will have good friends that you can count on.

- You will have people who care about you and are supportive.

- You will know people that you can have fun with.

- You will have a couple of good friends that you can talk to about personal things.

- You will share your life with someone you love.

Community Feelings

- You will work to make the world a better place.

- You will help others improve their lives.

- You will help people in need.

- You will donate time or money to charity.

- You will work for the betterment of society.

EXTRINSIC GOALS

Financial Success

- You will be financially successful.

- You will have a lot of expensive possessions.

- You will have a job with high social status.

- You will have a job that pays well.

Social Recognition

- Your name will be known by many people.

- You will be famous.

- You will be admired by many people.

- Your name will appear frequently in the media.

- You will do something that brings you much recognition.

Attractive Appearance

- You will have many people comment often about how attractive you look.

- You will keep up with fashions in hair and clothing.

- You will achieve the "look" you've been after.

- You will successfully hide the signs of aging.

- Your image will be one others find appealing.

Responses were recorded on a five-point scale. In responding to how important each aspiration was to them, participants selected an option ranging from 1 = "not at all important" to 5 = "very important." For their likelihood of attaining each aspiration, they could choose a response ranging from 1 = "very low" to 5 = "very high."

To measure the relationship between goal content and well-being, the investigators also included measures of self-actualization, vitality, depression, and anxiety—questions not listed above. The questionnaire was administered to a sample of 316 college students.

Using careful statistical analysis to study the proposed relationship between a person's goals (aspirations) and well-being, the researchers found that the initial study's results confirmed their suspicions. Compared to students who placed a higher importance on the goals of self-acceptance, affiliation, and community feeling, those who placed a higher importance on achieving financial success reported lower levels of self-actualization and vitality. Additionally, those who placed a higher importance on financial success also reported higher levels of depression and anxiety. In sum, a stronger focus on material aspirations was consistently related to lower levels of psychological well-being. However, the researchers emphasized that financial success per se is not bad; it's when that form of success is *valued* more than other life domains that it becomes a negative factor.

As noted earlier, in a second (two-part) study published in 1996, Kasser and Ryan—who by this point had recognized and focused on the distinction between intrinsic and extrinsic aspirations—expanded their Aspiration Index to include two additional extrinsic goals considered to be part and parcel of the consumer culture: social recognition and attractive appearance (image). Although striving for money and material possessions is at the core of the American Dream, being famous (social recognition) and being attractive are important goals and values of our current consumer culture.

To test the second project's basic premise that a relative focus on extrinsic goals would lead to lower well-being, two studies were conducted. In the first, 100 adults residing in Rochester, New York, completed a questionnaire that contained the newly expanded Aspiration Index as well as measures of self-actualization, vitality, depression, anxiety, and physical health (assessed by how often in the past seven days a respondent had been bothered by nine different physical ailments, including headaches, faintness, and sore muscles). Our love of stuff couldn't be related to how we physically feel, could it? Let's just see.

As hypothesized, a greater importance placed on the extrinsic goals of financial success, social recognition, and attractive appearance was associated with lower levels of vitality and self-actualization. Those with an extrinsic orientation also reported more headaches, backaches, sore muscles, and sore throats than those who expressed a more intrinsic orientation. This was one of the earliest indicators that materialistic values negatively impacted not only psychological well-being but also physical well-being.

In a second study, 192 college students completed a questionnaire that included the same questions used in the previous study, with the addition of a scale to measure narcissism and a daily checklist. The daily checklist required that the students answer several questions about their current situation. Twice a day, in the middle of the day and at night, the students recorded the extent to which they had each of the same nine physical ailments measured in the first study and also how often they felt each of nine different emotions—four positive moods (e.g., happy, joyful) and five negative moods (e.g., unhappy and angry).

Consistent with the results found in the first part of the first project, those participants who espoused a more extrinsic goal orientation were less self-actualized and vital and reported more depression than those who were more intrinsic in their goal orientation. Additionally, those with an extrinsic orientation reported more physical symptoms and reported fewer positive emotions over the two-week exercise. It appears that a strong desire for materialistic pursuits negatively impacts how we experience our day-to-day lives, as well as our longer-term physical and psychological well-being.

And, no great surprise here, those who expressed a more extrinsic orientation were also found to be more narcissistic. This is telling, because narcissism often manifests itself as an attempt to hide inner feelings of emptiness and low self-worth with a grandiose façade that screams, "Look at me: I'm a person of value." Being vain, narcissists desire special treatment and admiration from those around them. Many people claim that our current consumer culture encourages narcissism by extolling the virtues of consumption. Take, for instance, the popular tween retailer Justice, whose "Princess" tops and accessories are sold with the slogan "Love yourself." In Club Libby Lu, a spa that until recently was offered in some Saks stores, parties of five- to thirteen-year-old girls were given Princess makeovers, with—as Megan Basham of the *Wall Street Journal* puts it—"tube tops and miniskirts that left girls looking more like Real Housewives than Cinderella."[8] This and a slew of other ad campaigns and slogans focused on self-indulgence dovetail nicely with our current understanding of materialism. A narcissist's desire for attention from others is consistent with extrinsic pursuits whose primary purpose is to garner the approval of others.

Numerous studies have followed Kasser and Ryan's early work, and these later projects provide further support for the negative relationship between materialistic values and personal well-being. A study by Patricia Cohen and Jacob Cohen found that adolescents who expressed material values such as "having expensive possessions" and "wearing expensive clothes" were more likely than the typical adolescent to have mental disorders such as depression and paranoia.[9] A later study by Carol Nickerson and several coauthors found that people who'd had stronger financial aspirations in 1976 reported lower levels of life satisfaction nineteen years later, compared to those who were less pecuniary in their pursuits.[10]

Given this book's focus on consumerism, it makes sense to move from this clearly established link between extrinsic goals and well-being to look at whether such goals have an impact on our behavior as consumers. As a uniquely consumer value, one would expect that financial success, for example, would have a significant impact on how we shop and spend money. Let me give you a short glimpse (actually, a brief excerpt) of a study I conducted with my good friend Stephen Pirog of Seton Hall University.[11] The primary focus of that study was to investigate the role that extrinsic and intrinsic goals play in an all-too-common consequence of our consumer culture: compulsive buying. As discussed in chapter 5, compulsive buying can have severe consequences not only for those directly involved—the compulsive buyer and his or her family and friends—but also for the larger community. Much of the economic troubles we are experiencing as a nation today can be traced to spending money beyond our ability to pay.

The study's hypotheses were straightforward. We predicted that the three intrinsic goals of self-acceptance, affiliation, and community feeling would all be negatively associated with compulsive buying. Put another way, since compulsive buying is believed to be largely driven by negative thoughts and emotions, we expected that those who value intrinsic goals would be less likely to be compulsive buyers. On the other hand, we hypothesized that respondents who espoused the relative importance of the three extrinsic goals—financial success, social recognition, and an attractive appearance—would be more likely to exhibit compulsive buying behavior.

Our study found strong support for the argument that extrinsic goal systems are positively associated with compulsive buying, while intrinsic goal systems are negatively associated. In particular, two intrinsic goals (self-acceptance and community feeling) and two extrinsic goals (financial success and attractive appearance) were significantly correlated with compulsive buying. It may be that extrinsic goals are symptomatic of a more insecure personal style, and that certain people address these insecurities through purchasing, consuming, and displaying material possessions.

On the flip side, the pursuit of intrinsic goals appears to have a salubrious effect on one's well-being. The satisfaction that these goals bring to important psychological needs appears to buffer individuals oriented toward intrinsic goals from a consumer culture that extols material possessions as the path to happiness. The connection between intrinsic goals and compulsive buying may be self-esteem—that is, people oriented toward intrinsic goals may enjoy higher self-esteem and see little value in material possessions as a means to happiness.

A particularly interesting finding of this study has to do with gender differences and financial success. Men and women in the group placed an equal importance on financial success, but men saw their chances of obtaining it as better than did women. With a lower perceived likelihood of achieving financial success, women may focus more on the trappings of financial success than on financial success itself. Women also placed a higher importance than men on maintaining an attractive appearance. Earlier research found that female compulsive buyers reported higher expenditure levels on clothing and cosmetics.[12] This suggests that the desire for an attractive appearance may be the root cause of compulsive buying in women.

As this chapter has shown, the goals we hold as most important play an integral role in our well-being. Intrinsic goals such as self-acceptance, affiliation, and community involvement are nurturing and make us happy. Extrinsic goals such as financial success and materialism run counter to our happiness. We are a nation of people who profess family values (goals), but our behavior is not consistent with such sentiments. Actions always speak louder than words. Chapter 8 will now demonstrate how a contradiction between professed values and behavior can wreak havoc in one sure route to happiness—healthy relationships.

— 8 —

COLLATERAL DAMAGE: RELATIONSHIPS

There is only one happiness in life, to love and be loved.

—GEORGE SAND

My mother-in-law told me about a great T-shirt that read, "Please, Lord—give me a chance to prove that winning the lottery won't change me." I'm sure this is a sentiment we all share—win the lottery and all our troubles would simply fade away. Well, not quite. In a classic study on how winning the lottery impacts happiness, researchers Philip Brickman, Dan Coates, and Ronnie Janoff-Bulman interviewed twenty-two major winners of the Illinois state lottery. The trio also interviewed twenty-two average consumers and, interestingly, twenty-nine accident victims (including eleven paraplegic and eighteen quadriplegic respondents).[1]

Respondents in each of the three groups (lottery winners, average consumers, and accident victims) were asked to rate, on a five-point scale ranging from 0 = "not at all" to 5 = "very much," how happy they were now (not at this moment, but at this stage of their life), how happy they were before winning the lottery or having the accident (or six months ago, for the average consumers), and how happy they expected to be in a couple of years. The groups were also asked to rate how pleasant they found everyday events such as talking with friends, watching TV, eating breakfast, hearing a funny joke, getting a compliment, reading a magazine, and buying clothes.

Surprisingly, as the table below attests, lottery winners described themselves as being no happier than typical consumers assessing the past or the present, nor did they have any greater expectation of future happiness. The real kick in the pants is that lottery winners didn't foresee any greater future happiness than did the accident victims! When it came to enjoying the routine pleasures of life, lottery winners rated the seven everyday activities as less pleasurable than did the average consumer group, however. It appears that after a big event like winning the lottery, simple pleasures lose some of their luster.

The Happiness of Lottery Winners vs. Others

GROUP	GENERAL HAPPINESS PAST	PRESENT	FUTURE	EVERYDAY PLEASURES
Lottery Winners	3.77	4.00	4.20	3.33
Average Consumers	3.32	3.82	4.14	3.82
Accident Victims	4.41	2.96	4.32	3.48

All ratings were on a five-point scale that ranged from 0 ("not at all") to 5 ("very much").

Results from a study of lottery winners in the United Kingdom might help explain these surprising results. Although winning the lottery can be good, it also has its costs. Many UK lottery winners quit their jobs and moved to new (presumably fancier) neighborhoods, thus losing contact with many friends. Additionally, family and friends who wanted to share in the winners' good fortune were disappointed when they received less than they "deserved." Other research suggests that a sudden increase in income increases the likelihood of divorce. The potentially negative impact of winning the lottery on interpersonal relationships is an important reason why lottery winners are not happier than the rest of us.[2]

THE IMPORTANCE OF GOOD RELATIONSHIPS

Our need to belong may be our strongest motivation as humans. We spend an inordinate amount of time thinking about our current and hoped-for relationships with other people. Think of the joy we experience when we make a new friend or receive a compliment from someone we care for. Likewise, think of the heart-wrenching pain of a failed relationship. The extent of that joy and pain indicates to us that relationships are the key to our happiness. As author and psychologist Tim Kasser aptly states, "Our psychological health depends in part on whether we feel close and connected with other people, and on whether we can give and receive love, care and support."[3]

Recall from the previous chapter that personal relationships, along with self-acceptance and community involvement, are intrinsic aspirations rooted in our need for personal growth. Good relationships are inherently satisfying (that is, done for their own sake). The locus of evaluation for intrinsically motivated behavior comes from within oneself. Intrinsically driven people look less for the approval of others in deciding what to do; instead, they pursue those things that promote personal growth and that are, in themselves, gratifying.

Researcher Simone Pettigrew conducted thirty-six interviews across three countries (twelve per country)—Australia, the United States, and the United Kingdom—to explore what people perceived to be an ideal life.[4] These three countries were chosen because they share similar standings in GDP per capita (a measure of consumption), life expectancy, and education levels. The primary focus of the interviews was on people's priorities and values and the extent to which the consumption of goods and services ranked as important facets of the good life. The interviews began with broad, open-ended questions that invited the interviewees to talk about themselves and their lives in an unstructured fashion. The interviews were taped and carefully analyzed by the researcher.

Interestingly, instead of wishing for a big lottery win or great collections of material possessions, most people were pretty satisfied with their lives. Given the Western world's preoccupation with material possessions, it came as a great surprise that possessions weren't a prominent feature of the interviewees' ideal life. Instead, and relevant to our present discussion, the lion's share of concern was with the quality of relationships with family and friends. This was particularly true for the female interviewees. Most females not in a long-term relationship wished to be so, and those without children looked forward to the day they could start their own family. Most comments were about immediate family members, but extended family and friends were also seen as critical to happiness. Related to the desire for good relationships with family and friends was a yearning for more time to be with family—not an uncommon condition in today's consumer culture.

While money was mentioned frequently by interviewees, wealth was generally not seen as necessary—in fact, many interviewees stressed that they didn't need lots of money. People simply wanted enough financial assets to ensure that they and their loved ones could be free from financial worries. Some did mention specific benefits/products (such as travel, houses, and cars) that might help them be more happy, but Pettigrew notes that "such mentions were typically brief and delivered in a less heart-felt manner" than discussions about the desire for good relationships.[5]

Modern living conditions have not helped us maintain healthy social relationships. Consider that humans evolved in close-knit groups of somewhere between 50 and 200 people. In contrast, modern members of the species Homo sapiens generally live in large cities surrounded by thousands if not millions of other humans. Most Americans live an anonymous existence in isolated nuclear families with frail (or nonexistent) kin networks. Like so many of you, my family has scattered across the country in the hot pursuit of better jobs and promotions—a mobility that undermines the social support which at one time was provided by our extended kin networks. As evolutionary psychologist David Buss notes, "If psychological well-being is linked with having deep intimate contacts, being a valued member of an enduring social group, and being enmeshed in a network of extended kin, then the conditions of modern living seem designed to interfere with human happiness."[6]

Ed Diener and Martin Seligman, both of whom have been mentioned in previous chapters, collaborated on a study to uncover the characteristics of happy people.[7] Using a sample of 222 college students, the researchers identified the upper 10 percent of very happy people and the lowest 10 percent of very unhappy people. On a life satisfaction scale where scores could range from 5 to 35, the happiest 10 percent averaged about 30 on life satisfaction, whereas the unhappiest 10 percent scored an average of 16 on the same life satisfaction scale. Compared to unhappy people, the happiest 10 percent were more likely to (1) spend greater amounts of time socializing, (2) have good relationships with others, and (3) be more outgoing (extraverted).

Diener and Seligman concluded that the primary difference between happy and unhappy people lies in the area of social relationships. In a nutshell, happy people tend to have better social relationships. What this study tells us is that having a good social life is a necessary, but not entirely sufficient, component of well-being—in other words, a person may or may not be happy with good social relationships, but without good social relationships the chances of achieving happiness are remote.

MATERIALISM AS PART OF OUR OVERALL VALUE SYSTEM

As we have already discussed, materialism is an important life value for many people, but it's just one component of a broader value profile. Put more plainly, we have other values that interact with materialism and with each other, collaboratively impacting both our behavior and our happiness. Values such as materialism can be meaningfully understood only when viewed as part of that larger value system. Values aren't discrete entities; they're messy and overlapping, and they often contradict each other. And yet somehow this complex interplay of values dictates the decisions that guide our behavior.

The best attempt I know of to explain the relationship among life values was undertaken by researcher Shalom Schwartz, who developed what is commonly referred to as "Schwartz's Circumplex Model of Values," shown below.[8] As you can see, Schwartz identifies ten basic categories of values: power, achievement, hedonism, stimulation, self-direction, universalism, benevolence, conformity, tradition, and security. He didn't just think those values up: to arrive at them, he collected data in a series of studies from adults, college students, and teachers in forty countries across the globe.

Schwartz's Circumplex Model of Values

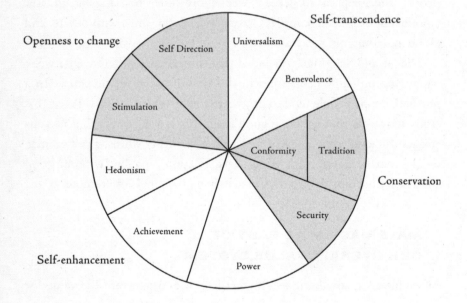

Definitions of Schwartz's General Value Types

- **Power.** Social status and prestige, control or dominance over people and resources.

- **Achievement.** Personal success through demonstrating competence according to social standards.

- **Hedonism.** Pleasure and sensuous gratification for oneself.

- **Stimulation.** Excitement, novelty, and challenge in life.

- **Self-direction.** Independent thought and action; choosing, creating, exploring.

- **Universalism.** Understanding, appreciation, and tolerance for all things (people and nature); a general concern for others.

- **Benevolence.** Preservation and enhancement of the welfare of people with whom one is in frequent personal contact.

- **Tradition.** Respect for, commitment to, and acceptance of the ideas and customs of the past.

- **Conformity.** Restraint of actions, inclinations, and impulses likely to upset or harm others or violate social expectations and norms.

- **Security.** Safety, harmony, and stability of society, of relationships, and of the self.

Participants in Schwartz's study were asked to rate the importance of a long list of values—many more than the ten listed above—which were then analyzed using a fancy statistical technique that located the values in a circular space—hence the wording "Circumplex of Values." If you take a close look at the circumplex, you will notice that values that ended up close to each other are ones that most people would consider to be compatible, whereas values at the opposite poles of the circle are ones that most people would find contradictory or conflicting. The overlap between the values is such that they share a "slice" of the values pie.

Schwartz developed this arrangement of values using two axes: (1) self-enhancement versus self-transcendence and (2) openness to change versus conservation. Schwartz did not include a measure of materialistic values but they are quite similar to the values of power, achievement, and hedonism, all located on the self-enhancement axis. The opposing axis, self-transcendence, focuses on other-oriented values, including benevolence and universalism.

Materialism's location in the circumplex is critical, because it explains how that value relates to other commonly held values. The pursuit of materialistic goals, according to Schwartz's circumplex, is in direct conflict with the values of benevolence and universalism, both of which have a strong social component—dealing with friends and family (benevolence), dealing with people and nature (universalism). Schwartz considers the juxtaposition of complementary (adjacent) versus conflicting (opposite) values to be the key premise of his model. Specifically, he is interested in the notion of "values conflict" as a source of psychological tension. It is not uncommon for someone to hold conflicting values such as hedonism and benevolence, or financial success and friends and family. The conflict produced by such opposing values is responsible for the negative relationship between materialism and happiness. Consider a man who, staying late at work in pursuit of an elusive promotion, loses out on tucking his young children into bed at night.

HOW MATERIALISM UNDERMINES RELATIONSHIPS

So what is it about materialism that undermines healthy relationships? At least four possible explanations exist:

- Materialists tend to "objectify" others around them; to a materialist, other people are used as a means to an end—their own happiness.

- Materialists are self-centered, meaning that they often crash and burn in the two-way street of relationships; maybe this is why materialists are more likely to get divorced than those who hold more self-transcendent values.

- Materialists place a low value on relationships; other people just aren't high on their list of priorities.

- Materialists tend to opt for work over family when pushed by time constraints; we're a nation of overworked souls, and the family is a major casualty.

Seeing People as a Means to an End

Thorstein Veblen, in his now classic book *The Theory of the Leisure Class*, studied the consumer behavior of the moneyed elite at the turn of the twentieth century. It was during this research that he coined the phrase "conspicuous consumption." He also noted that the rich of his day treated others as objects, commenting that captains of industry tended to collect what today we call "trophy wives" just as they might collect other consumer objects. Trophy wives were purchased (used) to convey status upon their "owner."[9]

While the subject of seeing people as chattel isn't new, we don't usually link that attitude to our role as consumers. However, the link is there—and it's strong. The consumer culture's objectification of humans is insidious and pervasive; it's by no means the purview only of the upper class these days. Like so many things, what started as a behavior exhibited by the relatively affluent has trickled down to the hoi polloi (that's you and me).

Simply put, materialism affects the way we interact with others. An emphasis on shopping, spending, consuming, and assigning monetary value to things bleeds over into our relationships. It leads us to treat others as objects; and when we see others as objects, we're less concerned about their experiences and feelings. Researchers Yiannis Gabriel and Tim Lang may have expressed it best:

> By the beginning of the 21st century, we have learned to talk and think of each other and of ourselves less as workers, citizens, parents or teachers, and more as consumers. Our rights and our

powers derive from our standing as consumers; our political choices are votes for those promising us the best deal as consumers; our enjoyment of life is almost synonymous with the quantities (and to a lesser extent quality) of what we consume. Our success is measured in terms of how well we are doing as consumers. Consumption is not just a means of fulfilling needs but permeates our social relationships, identities, perceptions and images.[10]

This "commoditization" of human life comes at the expense of our relationships. Consuming has become the template or model against which we view all of our activities—including relationships. As researcher Zygmunt Bauman puts it, consumerism has "become a kind of default philosophy for all modern life."[11] It's as if we, with our shiny-objects ethos, lack the imagination to fathom a life without material possessions at its center.

Materialists are experts at manipulation. Think about those "social climbers" we talked about earlier—people who choose friends to improve their social standing and bolster their self-esteem. The bonds forged from such relationships are fragile and fleeting; quickly dissolving at the first sign of trouble, or when the other person no longer fulfills the need for which he or she was "recruited." Barry Schwartz, who calls such relationships "instrumental friendships," says that in our current consumer culture, all that's required is that each "friend" gets something of value. Instrumental friends, unlike people who have mutual interests and genuinely care for one another, use each other to achieve certain goals. Schwartz argues that "instrumental friendships are very similar to legal contracts, with personal contact and the knowledge of mutual interdependence substituting for formal contractual documents."[12]

Shivani Khanna and Tim Kasser looked into this sort of friendship, testing a sample of college students from the United States, India, and Denmark. They found materialism to be strongly associated with the tendency to use others for instrumental purposes. Students with a strong inclination to objectify others agreed with statements such as "I like popular people," "I like to be with 'cool' people because it helps me look 'cool' too," and—maybe most telling—"If a friend can't help me get ahead in life, I usually end the friendship."[13]

"It's All About Me"

In addition to treating people as objects, those who are highly materialistic are also more self-centered than less materialistic people. Russell Belk, in pioneering work in the field of materialism, conceptualized materialism as consisting of three traits or characteristics: possessiveness, non-generosity, and envy.[14] Highly materialistic people prefer to own and keep things instead of renting or borrowing or discarding objects; they are not generous, preferring not to share with others; and they tend to be envious of what others have and to become upset when others have things they don't. All told, not a very pretty picture of materialistic people.

The survey that Belk used to study materialism was administered to a sample of people from a wide swath of society. In addition to asking twenty-four questions designed to measure the respondents' possessiveness, non-generosity, and envy, he also included two questions regarding participants' well-being: "How happy are you?" and "How satisfied are you with your life?" As you might have guessed, those who expressed high levels of materialism reported that they were less happy and satisfied with their lives than their less materialistic counterparts.

A study by Marsha Richins and Scott Dawson further established the link between materialism and the traits of selfishness and non-generosity.[15] These researchers asked a large sample of adults in the United States how they would spend an unexpected windfall of $20,000. Highly materialistic participants said that they would spend an average of $3,445 on "buying things I want or need." This figure is three times larger than the $1,106 that those low in materialism reported that they would spend on themselves. These same highly materialistic people were also less generous than their less materialistic counterparts. When asked how much money they would give to a local church or charity, less materialistic participants reported that they would give an average of $1,782, compared to $733 for their highly materialistic brethren. Friends and relatives were also casualties of materialism. Highly materialistic people earmarked only $1,089 of their $20,000 windfall to give or lend to friends and family, compared to $2,631 for less materialistic people.

But it gets worse: new research has found that spending money on gifts for friends or donating money to charity increases one's happiness, while spending money on oneself does not. This isn't good news for those of us pursuing happiness through shopping. Elizabeth Dunn, a psychologist at the University of British Columbia working with several colleagues, surveyed a nationally representative sample of 630 Americans.[16] Respondents answered questions about their happiness, income, and spending habits, including charitable donations and purchases made for themselves or for others.

The study's results revealed that personal spending had no link to one's happiness, while spending money on others or making a donation to charity was a reliable way to boost one's happiness. The researchers found that, regardless of income level, those who spent money on others reported greater happiness than those who spent more on themselves. In a related study, Dunn and her colleagues found that employees at a Boston company who devoted more of their profit-sharing bonus to others—you guessed it—reported higher levels of overall happiness than those who spent their bonuses on themselves.

In fact, spending even a small amount of money on others boosted happiness. The experimenters gave college students a $5 or $20 bill with the instruction to half that they should spend the money on themselves and to the other half that they should spend it on others. Those who spent the money on others—buying, for example, a small toy, a gift, or a shared meal—reported feeling happier at the end of the day than did those who spent the small windfall on themselves. It's amazing to think that spending even as little as $5 on someone else can boost your happiness! This is certainly convincing evidence that the selfishness and lack of generosity associated with materialism run counter to our happiness.

Most telling might be the results of another study of over 100 college students conducted by these same researchers. Of those polled, most thought that spending money on themselves—just *contemplating* spending money, not actually doing it, as in the prior experiment—would make them happier than spending on others or donating money to charity. It appears that at some level many of us really do believe that buying things will make us happy.

Seeking Personal Space

Ample evidence exists that materialists place a lower value on interpersonal relationships than those with more moderate views on the value of money and possessions. A series of particularly insightful and astonishing studies by researchers Kathleen Vohs, Nicole Mead, and Miranda Goode found that highly materialistic people prefer less social intimacy—particularly when money is involved.[17]

MATERIALISM AND MARITAL (UN)HAPPINESS

A recent study by Jason Carroll of Brigham Young University finds that highly materialistic spouses are about 40 percent more likely than less materialistic spouses to experience high levels of financial problems (surprise, surprise), which in turn lowers their marital satisfaction. The study's results are based on a nationally representative sample of 600 married couples, of which 35 percent reported high levels of materialism. Carroll's team observed that a materialistic spouse is a better predictor of a couple's financial problems than is the couple's income. Carroll concludes that materialism increases financial problems in marriage for two reasons: (1) a highly materialistic spouse may spend money irrationally chasing materialistic dreams, which causes money problems, and (2) highly materialistic spouses have a larger gap between their current and ideal states, which leads to more complaints and conflicts over money even in the absence of problems such as debt.[18]

Vohs and her colleagues—these are the same researchers who did the experiment with Monopoly money mentioned in chapter 1—conducted three experiments to test whether people who were reminded of money ("high-money condition") preferred more or less social contact than those who received no such reminders ("low-money condition"). In their first experiment, participants were seated at a desk with a computer. The only difference was the screensaver on the computer. The control group was exposed to a blank screen, while the low-money group saw a fish screensaver and the high-money group saw a screensaver with images of money.

After being seated at a desk, participants were told that their next task entailed getting acquainted with a participant in another room. As the experimenter left the room ostensibly to get the other participant (though there actually were no other participants), she directed the attention of the participant to a chair in the corner of the room, instructing the participant to pull that chair toward his or her own chair for the "get acquainted" exercise. The distance between the two chairs, measured in inches, was taken as a sign of preferred social intimacy. As hypothesized, those participants who were reminded of money allowed more space between the two chairs than either of the other two groups—blank and fish screensavers.

In a second experiment, the researchers tested the hypothesis that being reminded of money would prompt separation from strangers, but surely not from friends and family. To activate the concept of money, participants were asked to complete questionnaires while seated at a desk that had either a poster of money—a very subtle reminder—or a pretty watercolor poster within their visual periphery. After completing the questionnaire, participants were given a list of leisure activities and asked to indicate which they thought were most enjoyable. The list was designed in such a way that participants were forced to choose between solitary activities (for example, reading a book) or shared activities. To test the idea that friends and family would triumph over money's tendency to encourage social separateness, the researchers included activities that specifically mentioned friends, family, and loved ones. Even when given the opportunity (on paper) to spend time with close friends and family, those in the high-money condition chose solitary leisure activities more often than did the low-money participants (those exposed to the watercolor poster).

This experiment is a particularly revealing study of how values such as materialism crowd out the more nurturing pursuits of friends and family. Material values are not, in and of themselves, bad, as noted earlier. It's when such values are held in higher esteem than personal relationships— as the above-mentioned study suggests they often are—that they undermine our happiness.

What's interesting in all of this is that the family seems to be partially to blame for its materialistic children. As an important socialization agent, the family—if it doesn't do its job well—can produce insecure offspring that cling to possessions as a source of security and self-esteem. In several studies, how nurturing mothers were with their children was related to their children's level of materialism. When mothers were more supportive of their children and provided them a secure environment in which to develop, the children expressed lower levels of materialism.

FAMILY FEUD   The headline of the *Daily Mail* (United Kingdom) story "Destroyed by a Shopaholic Daughter" nearly says it all. Norman and Carol Bishop were doting parents who were patient and generous (possibly to a fault) with their children. It was beyond their imagination that their eldest daughter, Laura, would take such advantage of their kindness and generosity. After all, they were (emphasis here) family.

The problem started around 1992, when Laura— unbeknownst to her mother—ordered £800 (multiply that amount by approximately 1.5 for American dollars) of stuff from various catalogs, using her mother's credit card but without repaying her mother. Things spiraled quickly downward after that.

The following year Laura and her husband, Steve, moved in with the Bishops after Laura's flat was repossessed due to mortgage problems, and they stayed a rather protracted three years. Both Laura and her husband worked full-time jobs, but they never offered a penny for rent or living expenses—although there appeared to be ample money available for shopping sprees. Even the windfall of a £50,000 inheritance from her grandmother didn't appear to help Laura's situation; she was soon short of cash again. Somehow, despite her record with money, Laura convinced her parents to take out a second mortgage on their home so that they could loan her £35,000. The money from the ill-advised second loan appeared to vanish into thin air. Laura never repaid a penny of it.

Believe it or not, several years later Laura convinced her parents that she needed more money and they should help out. With the best intentions, Norman gave his daughter his credit card with instructions to withdraw only £500. When the credit card bill arrived at the Bishops' now heavily mortgaged home, it revealed that their daughter had spent £9,000 on various items, including several large cash advances. Mr. Bishop, a retired electrician, had to come out of retirement and work fifty-five hours a week as a taxi driver to stay afloat.

As you might imagine, there's been considerable friction in the family over these various loans, and all the arguing has taken its toll. Norman's blood pressure has been one casualty of all the fighting. By far, though, the relational costs have been the hardest to endure. The once close-knit family has disowned their profligate daughter.[19]

In this regard, it's mothers, not parents more generally, that seem to be the key. Research by Eirini Flouri, using data from 2,218 British secondary school–age children, found that the father's involvement in child-rearing had no impact on a child's materialism but that the mother's involvement lessened the likelihood that the child would be materialistic. The mother's material values, however, were positively related to the child's material values—more materialistic mothers begat more materialistic children. Furthermore, materialistic adolescents, in general, reported lower levels of satisfaction with their relationship to their parents.[20]

The above suggests that parents who weren't as nurturing as they might have been with their youngsters, and who didn't do as much as they could have done to foster secure, well-adjusted offspring, sometimes raise insecure teens who use material possessions in an attempt to fill the hole left by their family environment. Tim Kasser, whose leading research in the field of materialism and well-being has already appeared in these pages, sees two reasons why such teens turn to materialism. First, teens who are insecure may be more susceptible to the daily marketing on-slaught whose primary job is to foster insecurities and then provide con-sumer remedies for such ailments—real or imagined. Second, they may turn to materialistic pursuits as a way to gain approval from others, which in turn would help them feel good about themselves. In a consumer cul-ture that encourages the pursuit of happiness through possessions, it makes sense that teens would turn to possessions in an attempt to gain the approval of others.[21]

It's not just the kids, though. Compelling research by the duo of Paul Amato and Stacy Rogers finds that material longings can undermine mar-ital bliss.[22] Starting with a sample of over 2,033 married couples under the age of fifty-five, Amato and Rogers investigated the extent to which re-ported marital problems in 1980 predicted divorce between 1980 and 1992. Respondents were called in 1983 and 1992 and asked the following:

> Have you had a problem in your marriage because one of you (a) gets angry easily, (b) has feelings that are easily hurt, (c) is jeal-ous, (d) is domineering, (e) is critical, (f) is moody, (g) won't talk to the other, (h) has had a sexual relationship with someone else, (i) has irritating habits, (j) is not home enough, (k) spends money foolishly, and (l) drinks or uses drugs?

Given that this chapter is about materialism and its impact on rela-tionships, we'll focus here on letter k above. Somewhat to my surprise, husbands more often than wives were said to "spend money foolishly" (11.5 percent versus 7.5 percent). Most shocking was the researchers' finding that the marital problem of spending money foolishly had a devastating impact on marriages. Couples who reported that one or both spouses spent money foolishly increased their odds of divorce by 45 percent! This is consistent with the research I reported on earlier which found that highly materialistic spouses were about 40 percent more likely than non-materialistic spouses to report high levels of financial problems, which in turn lowered their marital satisfaction. No doubt about it: relationships suffer when extrinsic values such as financial success are prized too highly.

Waiving "Quality Time"

Time is one of the most influential, mystifying, and often frustrating entities of our modern world. Some have touted it as our most precious resource, since we can never seem to consume enough of it. For many people, time is more valuable than even money—literally. Think, for example, of someone who negotiates a lower salary to gain more time off. If someone doesn't use time wisely, that particular time is gone forever, whereas money typically remains available for use at a later date.

It is not surprising, therefore, that a person's perception of time, and of how much or little there is, can influence well-being. As a result of what some social scientists have labeled the "time-compression effect," modern society is rife with labor-saving devices and efficiency measures. Despite this, we seem to have less and less time. This sense of a time-compressed lifestyle can impact your sense of well-being and, more relevant to our current discussion, can weaken your relationships. It was William Penn who long ago said, "Time is what we want most, but . . . what we use worst."

"Time affluence" can be thought of as how rushed, hectic, or busy one feels. High levels of time affluence suggest more "free" or "spare" time (and accordingly less time-caused stress). Juliet Schor, previously quoted author of *The Overworked American* and *The Overspent American*, claims that one of the most startling aspects of American society is how much we work. Annual work hours in the United States have increased from 1,716 in 1967 to 1,878 in 2000. Compared to other economically developed nations such as Norway and the Netherlands, Americans work an additional ten to fourteen weeks each year (approximately 500 hours). Specifically, notes author and activist Barbara Brandt, the average American male worked an average of 49.4 hours each week in 2000, while females worked an average of 42.4 hours each week. Both men and women alike reported that they would prefer to work some eleven hours less each week. Additionally, Americans reported wanting more free time— approximately two-thirds felt too much pressure to work longer hours.[23]

All that physical absence from home, and its resulting emotional detachment, leaves men and women impatient and ill-equipped to deal with the inevitable conflicts, big and small, which being a family entails. As family bonds weaken, family members drift apart and relationships and entire marriages dissolve. I don't believe it's happenstance that divorce rates are quite high in consumer cultures. We're simply too overburdened from work and consumer pursuits to invest the energy and time needed to maintain healthy marriages and families.

You might remember from our discussion in chapter 3 that Juliet Schor has identified a strong connection between long work hours and consumerism—a connection she calls the "work-spend cycle." The desire for material possessions drives consumers to work longer hours to afford the consumer goods they desire. I like researcher Annie Leonard's extension of Schor's work-spend cycle to include watching TV—a nearly universal pastime in the United States. For Leonard, the work-spend cycle becomes the work–watch TV–spend cycle. After a long day at work, we plop ourselves down in front of the television, where we're bombarded with advertising messages and depictions of a lavish lifestyle that leave us yearning for more. Not even mealtime is sacred anymore. Did you know that approximately two-thirds of American families watch TV while they eat? So much for quality "family time"![24]

As we saw earlier, all this TV viewing means that we're awash in advertising singing the virtues of consumption and promising that happiness is only a purchase or two away. To buy all of these "needed" items, we need money, however. And, unless we have a rich uncle, more money requires that we work longer hours. Being away from home so much, we try to make up for our absence by giving family members gifts that cost money. Author Arlie Hochschild refers to this process—one that repeats itself endlessly—as the "materialization of love."[25]

In addition to trading gifts for time and attention, adults have also imposed their frenetic, overworked lifestyles on their children. Young people today are left with little free time. Since the late 1970s, reports the University of Michigan's survey research center, U.S. children have lost twelve hours of free time each week and have little time for unstructured play. U.S. children are chronically sleep-deprived from getting up too early, going to bed too late, and having too many competitive activities scheduled.[26]

How we spend our time plays an important role in our happiness. As we saw in chapter 1, happiness researcher Sonja Lyubomirsky suggests that engaging in experiential activities with family members accounts for approximately 50 percent of our well-being. And yet materialism would pull us away from those important people. The revealing story of how we came to be so materialistic, at our own peril, is the subject of the next three chapters.

WHY ARE WE SO MATERIALISTIC?

Money never made a man happy yet, nor will it. The more a man has, the more he wants. Instead of filling a vacuum, it makes one.

—BENJAMIN FRANKLIN

It appears that when we feel threatened, either economically or by reminders of our mortality, we turn to money and possessions as a source of security. Let's look at our response to economic threat first; then, later in this chapter, we will look at the threat of mortality.

In a fascinating study, Kennon Sheldon and Tim Kasser exposed 447 college students to one of two scenarios. Of that group, 249 students were asked to read a short passage that talked about all of the current economic upheaval and the prospects of finding a job after college in such turbulent economic times—overall, a very bleak picture of the students' job prospects ("threat condition"). The luckier 198 participants read a much rosier narrative on the economy and their prospects for finding a job ("no-threat condition"). Both before and after that reading, they were asked questions about their goal orientation.

The results of this experiment revealed that participants in the no-threat condition did not change their goal orientation. In contrast, those participants who were made to feel anxious about their economic future (threat condition) gravitated toward extrinsic goals such as financial success and away from intrinsic goals such as self-improvement, friendships, and community involvement. When threatened economically, then, we tend to revert to a more primordial existence.[1] As we will see in an upcoming section, reminders of our mortality also bring out our materialistic impulses.

After all this talk about materialism, a good question to ask might be, How did we get to be so materialistic? In this chapter we will discuss two primary pathways to the development of materialistic values: (1) our materialistic culture, which creates (and then exploits!) feelings of insecurity, and (2) a genetic predisposition to consume.

PERSUADED TO CONSUME

The first pathway to materialism is exposure to social models that encourage materialistic values. People have a fundamental tendency to adopt the cultural values and behaviors that they see around them. That process of (usually unconscious) adoption is called "internalization," as we saw earlier.

Think about what we've learned in previous chapters: In our consumer culture, economic indicators are the primary barometer of quality of life. Money and material possessions are seen as ends in themselves, not the means by which we can travel or relax on the beach with our family. And the marketing machine used to encourage such consumer values is vast. We use goods to define ourselves and our aspirations, and we seek the meaning of life in stuff. Our economy requires that we buy, consume, discard, and replace goods and services at an ever-increasing rate to keep its engines chugging along.

Socialization Agents

Trying to instill healthy attitudes about money and materialism in our children conjures up the image of David (you) and Goliath (Madison

Avenue). In this story, however, David does not slay the giant; Goliath is just too powerful. Some experts claim that we're exposed to several thousand messages daily from Madison Avenue. When you consider all of the potential sources of marketing messages—branded clothing (including caps and shoes), billboards, radio, TV, Internet, cell phones, direct mail, storefronts, product placements in TV shows and movies—it's surprising that the number isn't higher. Just close your eyes and think of a simple trip to the grocery store; think of the products and images that assault you there.

If I asked you whether you'd like to spend the next five months doing nothing but watching television, surfing the Internet, and listening to music, you might think that's a little bit too much of a "good" thing. (Your teenagers might not, but you probably would. Who would want to spend the next five months doing that?) And yet, believe it or not, recent data from the U.S. Census Bureau tells us that adults and teens will spend nearly five months (3,518 hours) next year watching TV, surfing the Internet, and listening to personal music devices! It makes you stop and think about the impact such media exposure has on a person—especially young people, who are being exposed to media from the time they are in diapers. The average kid between eight and eighteen years of age spends nearly seven hours every day plugged into some type of media.[2]

Television is an important socialization agent and a major conduit of materialistic values. This should come as no surprise, given Americans' love affair with TV. Americans top the TV-watching charts, logging a staggering 1,820 hours per year on average (or approximately 2.5 months of nonstop viewing per year). The average child attends school for about half as long—roughly 900 hours every year. That same kid watches about 20,000 thirty-second commercials each and every year![3]

Does all that TV affect us? You bet! Research shows that heavy TV viewers are more materialistic than light TV viewers. In a study of 360 children between eight and twelve years of age, researchers Moniek Buijzen and Patti Valkenburg found that children who watched more TV asked for more products from their parents and reported higher levels of materialism. More disturbing was the finding that increased exposure to television advertising was also positively related to family conflict, increased disappointment after denied product requests, and greater life dissatisfaction.[4] Furthermore, because TV viewers use information from television to construct their views of the real world, heavy TV watchers have a distorted perspective of their surroundings, believing that others enjoy a higher standard of living than they really do and that luxury and status items are more commonplace than they really are.[5]

A seemingly simple solution to TV's deleterious effects would be to turn it off, right? No small task in our TV-obsessed culture. But it's not as "simple" as turning off the television anyway. That would ignore the important influence of family and friends on people's levels of materialism. For example, as we saw in the previous chapter, materialistic mothers are more likely to have materialistic children. That certainly highlights the importance of family as a socialization agent. Additionally, as research I was involved with found, teenagers who are more susceptible to the influence of their friends are more materialistic than their less susceptible counterparts.[6] Those of you with teenagers in the house know how strong that influence—peer pressure—can be. With the advent of two-wage-earner families and high levels of divorce, parents are playing a less active role as socialization agents. Peers have stepped in to fill this gap with less than stellar results. One result that's particularly relevant to our discussion here is that teens are more likely to purchase products when they're with their friends than with their parents.

THE GOOD LIFE   Jim Peyton, age sixty-one, of Shady Grove, Oregon (population 2,307), sports forty-one tattoos on his body as part of his cartoon character collection. Jim and  his wife, Karen, have surrounded themselves with thousands of objects festooned with Mickey Mouse, Sylvester the Cat, Snoopy, Daffy Duck, Speedy Gonzales, and other cartoon characters. The Peytons are moving full speed ahead with their collection, which includes cartoon-themed stuffed animals and figures, coffee mugs, refrigerator magnets, toys, T-shirts—you name it. They built an extra room on their house just to store their ever-growing collection, but they've already filled it with 500 cartoon cookie jars and countless other treasures.[7]

You're Okay—Am I Okay?

Psychology's "social comparison theory" (SCT) may help us to understand how all this works. SCT suggests that we come to know ourselves by comparing ourselves to others. We use how others look, dress, act, and consume as a crucial basis of comparison. Teens are particularly sensitive to this, looking to their peers for acceptance and to role models as they attempt to develop a self-identity. You know as well as I do that we all look to other people as barometers of our own success and worthiness. In essence, we form our self-concept based upon perceived appraisals by others. Although what we're hoping for is approval from others, upward social comparisons—that is, those where we look to people we perceive to be better than us—can engender feelings of inadequacy.

Advertising presents a nearly unlimited stream of opportunities for upward social comparisons. The models in the ads we see are always better looking, better dressed, richer, and more fun-loving than we are. But not to worry: there's a simple solution to narrow the social comparison gap—buy the product. This is particularly relevant given that materialists are more concerned than others with social comparisons.

The upward social comparisons that advertising relies on work so well because of human feelings of insecurity. And don't think for a second that we marketers don't know just how insecure most people are. In fact, many marketing appeals seek specifically to *foster* insecurity. Does my breath smell? Do I have BO? Does my car tell the world how worthwhile an individual I am? You get the picture: if we can foster a feeling of insecurity or self-doubt, a sale is soon to follow.

Materialistic consumption is a way that people attempt to compensate for worries and doubts about self-worth, their ability to cope, and issues concerning personal safety. That's why, as we saw in the previous chapter, less nurturing parents raise more materialistic children. Those poorly nurtured kids are more insecure and thus more susceptible to consumer messages that foster insecurities and then conveniently offer products to assuage such perceived inadequacies.

Although we're pretty universally insecure about how we compare with others, we've got even more important things to worry about. The biggest threat looming ahead of us is death. Research into attitudes about mortality has established a causal flow from feelings of insecurity to materialism. Subjects in such research are commonly primed to think of their own mortality with open-ended prompts about their death: "Describe the emotions that the thought of your own death arouses in you," for example, or "Jot down, as specifically as you can, what you think will happen to you as you physically die." The control group, on the other hand, typically responds to prompts about pain or other subjects not related to their mortality.

Several studies have found that manipulating "mortality salience"—in other words, reminding people of their mortality—increases the subjects' level of reported materialism.[8] We're reminded of that mortality (and scores of lesser anxieties) every time we read a paper or watch the news. Like those research subjects, we respond by embracing the cornerstone American value of materialism in an attempt to restore a sense of psychological security.

Dreams of Death

Materialism even follows us into the bedroom, but not in the way you might think. Research by husband and wife Tim and Virginia Kasser has found that those high in materialism have different types of dreams than less materialistic people.[9] Although some debate surrounds the interpretation of dreams, enough serious research has been conducted that we can say with confidence that our dreams reveal important insights into who we are, what concerns and conflicts we're dealing with, and what motivates our waking behavior.

The Kassers interviewed college students who scored either in the top 10 percent ("high materialists") or bottom 10 percent ("low materialists") on a scale designed to measure materialism. These students were asked to describe the two most powerful, meaningful, or memorable dreams they could recall. In looking for themes related to insecurity, the Kassers found three compelling differences between the dreams of the high materialists and those of the low materialists.

The first difference was that death played a more prominent role in the dreams of highly materialistic participants. Death was mentioned in a dream or in associations with a dream in 20.5 percent of the high materialists' dreams. In contrast, only 3 percent of the low materialists mentioned any association with death in their dreams. High materialists saw dead people (as if living) in their dreams and even ghostly figures dressed in black beckoning them from the grave.

The Kassers also found that 15 percent of high materialists' dreams involved a sense of falling, compared to only 3 percent in the low-materialism group. Kasser concludes, "Falling is almost universally interpreted by theorists as representing insecurity, as one is out of control, headed downward, and has nothing to hold on to."[10]

The third difference, a telling one, lay in the two groups' attitudes toward feared objects. Those low in materialism reframed feared objects so that they were no longer so frightening—for example, "That big scary dog just wanted to play with me." No one in the high-materialism group coped with fear in this manner. This difference in dreams suggests that people low in materialism may be better able to cope with their insecurities than highly materialistic individuals.

PROGRAMMED TO CONSUME

Another primary pathway to materialism, often overlooked, is that materialism may be part of our DNA. The relative roles of nature (genetics) and nurture (environment) as determinants of human behavior have been the subject of debate and controversy for hundreds of years. Neuroscientists and neuropsychologists believe that approximately 50 percent of human behavior can be explained by our genetic makeup.[11]

Evolutionary psychology is a new field of study whose aim is to unify the disciplines of evolutionary biology and cognitive psychology. Proponents of evolutionary psychology argue that human behavior cannot be fully understood without integrating knowledge of the natural sciences with that of the social sciences and humanities. In contrast to the traditional psychological paradigm, the emphasis of evolutionary psychology is on the *ultimate* rather than *proximate* explanations of our behavior as humans—in other words, on the biological source rather than closer-to-the-surface traits like happiness. Although much research remains to be done—we are really just starting to understand how the human brain works—it would be foolhardy not to consider a genetic explanation for our behavior as consumers. A possible genetic component to our behavior would by no means let us off the hook for our profligate spending, but it would help us understand why some of us have a tougher time than others controlling our materialistic longings.

A good example of the confluence of genetics and materialism can be found in the work of evolutionary psychologist David Buss.[12] Take as a given, for a moment, the fact that physical attractiveness is an important concern among women. Socialization and cultural pressures have been identified as key sources for this preoccupation with physical attractiveness. An evolutionary argument might be more appropriate, however—and here's where we get to Buss's work. Buss surveyed 10,000 people across thirty-seven cultures to determine the preferred characteristics of potential mates among men and women. His results are enlightening: across all (yes, *all*) cultures surveyed, men universally rated youth and looks as more important than did women, whereas in most cultures women preferred men with a higher earning capacity who were more ambitious and industrious. This is where materialism comes into the picture. Women desired men with a lot of stuff and men wanted a lot of stuff to attract women—preferences that maximize the likelihood of reproductive success ("Darwinian fitness") for both men and women and thus would have been the preferences in our evolutionary past. All very primal, no?

It appears that "evolution has supplied us with a highly developed 'on' button that drives us to seek more and more resources," notes Bob Cummins, professor of psychology at Deakin University.[13] This desire for more is useful only in a resource-strapped environment, however; thus it is not well suited for the consumer culture. We are particularly susceptible to such primal programming when it comes to money and material possessions.

There's further evidence of the importance of genetics to our behavior as consumers. Studies of identical twins have found a strong genetic influence for general intelligence and for personality traits such as aggressiveness and conscientiousness. Other research provides telling insight into the relative role of family socialization and genetics on behavior. Basic personality traits such as extraversion, agreeableness, conscientiousness, neuroticism, and openness to new experiences were found to have a genetic component—that is, they were passed from generation to generation. Surprisingly, sharing a common environment (living under one roof) had little or no influence on the same five personality traits! This suggests that a shared home environment has little effect on human behavior in general, and presumably on consumer behavior as well.[14]

Some go so far as to dismiss entirely the role of nurture. I'm not ready to go that far, but recent advances in neuroscience research require that we acknowledge the possibility of an increased role for genetics as determinants of human behavior. A leading scientist, in summarizing the state of nature/nurture research in a recent *Wall Street Journal* column, explained the role of nurture in the following manner: polishing up the things our children already do well and smoothing out the rough edges of the things they don't do so well.[15] Anyone who is a parent must acknowledge that there's a lot of truth in that statement. My two children have been physically and temperamentally different from each other since birth, and I bet yours have been as well. Our eldest is highly sociable (we call her "aggressively friendly"), while our youngest is quietly confident, likes time alone, and has a different build. These differences have been consistent until now—they're in their preteen years—and I have no expectation that they will change throughout their adult lives.

Interestingly, however, research in this area tells us that the role of genetics changes over one's life span. As I tell my marketing students, most of whom are twenty or twenty-one, they are probably as different from their parents now as they ever will be. As they continue around the circle of life, gradually aging, they will slowly become more and more like their parents. For example, general cognitive ability is only 50 percent genetics during adolescence, but it increases to 80 percent by the age of sixty-five.[16] This might explain why we start to think, talk, and physically resemble our parents as we age!

In summarizing the research on the role of genetics on adult personality, Martin Seligman concludes: "Studies find massive effects of genes on adult personality and only negligible effects on any particular events."[17] Researchers have found, for example, that adopted children share more in common as adults with their biological parents than with their adoptive parents. Similarly, notes Seligman, identical twins reared apart are far more similar as adults than fraternal twins reared together across a broad spectrum of personality types. You've heard stories on the news, haven't you, about identical twins raised apart who later find each other both, say, working at a Walmart and sharing eerily similar life experiences and spouses? For some of you, the importance of genes to adult personality and behavior may be a comforting thing. For me, though, thinking about my gene pool keeps me up at night.

In John Mowen's Model of Motivation and Personality, materialism is included as an "elemental" personality trait, along with emotional stability, introversion, conscientiousness, openness to experience, agreeability, need for arousal, and body focus.[18] Materialism, defined as the need to collect and possess material things, is seen by Mowen as one of the eight basic building blocks of personality, shaped by genes and early childhood experiences. These elemental traits impact our behavior directly and indirectly. Materialism's impact may directly influence our consumer behavior by increasing our desire to acquire and possess material objects. It may also affect our behavior as consumers through its impact on such traits as self-esteem and impulsiveness, both of which have been found to be correlated with materialism across many research studies.

A second study I conducted with Stephen Pirog supports materialism's impact on our consumer behavior—in this case, the behavior of credit card abuse.[19] In a survey of 254 college students, we found that materialism impacted the students' impulsiveness, which in turn increased the likelihood of student credit card abuse. This is an important finding, because our ability to control our behavior—in other words, to handle our impulsiveness—has genetic roots. Whether or not we act on our material longings appears to be the result of a complex interplay, then, between environment and genetic makeup. While a compelling case can be made for the consumer culture's impact on material values, biology—through its impact on our ability to exercise self-control—affects whether we succumb to consumerism and become a materialist.

In regard to consumer behavior, the most important site of influence is the human brain, a complex organ composed of approximately 100 billion neurons, a class of cells specifically designed for intercellular communication. Neurons come in a myriad of shapes and sizes that can result in a variety of patterns of interaction. Your genes and central nervous system largely determine the basic shapes and patterns of *your* particular neurons. The intercellular activity explains how genetics can influence our behavior as consumers.

Neurotransmitters, chemicals that allow the movement of information from one neuron to another, each have distinct chemical properties. Probably the most relevant neurotransmitters are monoamine derivatives of amino acids—namely, dopamine, norepinephrine, and epinephrine. A fourth neurotransmitter is serotonin, which has been used to treat compulsive buying with some success. These four monoamines are of particular relevance to consumer behavior because they are thought to influence our emotions and feelings of fear and pleasure. Endorphins, another class of neurotransmitters you may have heard of, are also linked to consumer behavior: they play an important role in the perception of pain.[20]

Researchers have identified a gene defect that restricts the activity of

MAO-A (one form of monoamine oxidase)—a digestive enzyme that inhibits the breaking down of two important regulators of human behavior, serotonin and norepinephrine. Without adequate MAO-A activity, people display a decreased ability to control impulses, which can lead to materialistic spending, impulse buying, and compulsive buying.

The impact of MAO-B (the brother to MAO-A) on our behavior is also commonly accepted. MAO-B functions by oxidizing dopamine, a neurotransmitter that affects your arousal level. Babies born with low levels of MAO-B are more aroused, active, and reactive than those with high levels—character traits that are emblematic of their behavior throughout their lives.[21] An individual's level of MAO-B is considered the biological force behind his or her level of "sensation-seeking"—that is, his or her desire for novel, varied, and thrilling sensations and experiences, and his or her willingness to take risks to experience thrills.[22] Sensation-seekers search out novelty in an unconscious attempt to increase their dopamine levels.

Are you a thrill-seeker, always living for new and different experiences? Do you like to take risks? Or are you a play-it-safe person who likes the safety of routine? Chances are you fall somewhere between the two extremes of the sensation-seeking continuum—not a total wild person, but not a stick-in-the-mud either. Whatever your sensation-seeking level, it's largely a function of your genetic makeup, with environmental differences playing little or no part. Marvin Zuckerman, a leading researcher in the area of sensation-seeking, has concluded that it's entirely genetic in origin.[23]

What does all this have to do with consumerism? A lot. Materialism has been found to be positively related to sensation-seeking. In other words, materialistic people get a thrill or "high" from buying possessions. Unfortunately, the thrill is short-lived and must be repeated again and again to achieve the desired pleasure state. This is precisely why we marketers work so hard to make product packaging appealing and to design stores that invite and excite shoppers. Our goal is to create the euphoric moments so many shoppers are after. These "rewards" are what keep shoppers coming back for more.

Our brains love novelty. Some of us, though, exhibit above-average levels of reward-seeking due to genetic influences. Excessive reward-seeking can lead to self-destructive consumer behavior such as impulse purchasing and compulsive buying, resulting in burdensome credit card debt; feelings of shame, guilt, anxiety, and depression; and interpersonal conflict with family members and other loved ones—not to mention the impact such behavior can have on society and the natural environment. Once again, materialism's negative impact on the quality of life is evident.

Although the nature/nurture debate is far from settled, it appears that in the consumer realm the individual is not entirely to blame. Nonetheless, there's still ample room for the non-genetic explanations discussed earlier in this chapter. Shopping may be part of our DNA, but that's not a sufficient excuse for overspending and consumer excess. It's only when we consider both sides of the nature versus nurture debate that we can begin to gain a glimmer of insight into our behavior as consumers.

Does God want you to prosper financially? In the next chapter we will investigate the "prosperity gospel" movement and the growth of mega-churches. Has the consumer culture leached into our places of worship, thereby sanctifying the pursuit of material possessions? Will a more devout spiritual life bring you financial blessings? Some say it will.

HEAVEN HELP US: THE PROSPERITY GOSPEL

If you want "pie-in the sky when you die," then Reverend Ike is not your man. If you want your pie now, with ice cream on top, then see and hear Reverend Ike on TV.

—AD FOR *THE JOY OF LIVING*,
HOSTED BY REVEREND IKE

The current economic crisis has all of us carefully scrutinizing the comings and goings of our money—even at church. Despite the bad economy, or maybe because of it, a theology called the "prosperity gospel" (PG) is gaining hold in many churches. Simply put, the prosperity gospel says that tithes and donations given to a church will result in financial blessings for the giver. I will have much more to say about this movement soon. For now, please take the somewhat tongue-in-cheek "PG Self-Examination Quiz" developed by Jonathan L. Walton, professor of religion at the University of California at Riverside.[1] The quiz will help you determine, in Walton's words, "whether or not you are a member of a prosperity gospel church and whether your pastor is a prosperity teacher." Have fun!

Prosperity Gospel Self-Examination Quiz

1. Does the name of your congregation end not in "Church" but in "Center" (i.e., Magnolia Christian Center as opposed to Magnolia Christian Church)?

   1 point

2. Does your pastor use language such as "seedtime and harvest" to describe the act of tithing?

   1 point

3. Does your church teach that God wants you to be a millionaire, yet your pastor is the only millionaire in the congregation?

   1 point

4. Does your pastor regard the current economic crisis as a "test from God," and thus has encouraged you to increase your financial contributions to the ministry in recent months as a "test of your faith"?

   1 point

5. Does your congregation meet in a multi-million-dollar state-of-the-art facility and talk about changing the world, yet the only other productive businesses within a five-mile radius are a Church's Fried Chicken, an AutoZone, and a payday lender?

   1 point

6. Does your pastor own any of the following: Rolls Royce, Bentley, Learjet, or helicopter?

   1 point

7. If you checked "yes" to the previous question, was it presented to the pastor as a gift from the congregation?

   3 points

8. Has your pastor earned an honorary doctorate from Oral Roberts University?

1 point

9. Did he/she sit on the board of regents at Oral Roberts University?

5 points

10. Did he/she resign from the board of regents at Oral Roberts University in recent years related to Richard Roberts's financial scandal?

10 points

11. Does your pastor teach that anyone who speaks out against prosperity teachings has a poverty spirit?

7 points

12. Does your pastor's name remind you of money (i.e., dollar, price, big money, etc.)?

7 points

RESULTS KEY

1–3 points—It is safe to say that you do not belong to a prosperity gospel ministry. Your pastor just may be a little ostentatious and/or have self-indulgent tendencies. We all have our moments . . .

4–6 points—Your church may not be a part of the Word of Faith movement, but the pastor spends less time preparing his/her sermon using Bible commentaries and more time watching TBN.

7–10 points—Yes, your church is a part of the Word of Faith movement and your pastor is a prosperity preacher. Who am I to judge? It just might work for you right now. But please be sure to protect your Social Security number, and don't hand over your W-2s.

10+ points—Sure, you have already handed over your Social Security number; your W-2 forms will be next in the offering plate.

Do you consider yourself a religious person? Do you attend a place of worship regularly? Pray? The answer is most likely yes: Americans, by and large, are a deeply religious people. And yet, visit one of the many houses of worship that dot the local landscape and you will experience the far-reaching impact of materialism. The boundaries between the secular and nonsecular worlds have all but disappeared in many churches today.

The prosperity gospel, often described as "name it and claim it theology," preaches that an authentic religious belief and behavior, usually in the form of tithes and other monetary donations, will result in material prosperity. Conversely, the gospel contends that financial prosperity and success in your private and professional lives is evidence of God's favor.

It should come as no big surprise that the theology behind the prosperity gospel was birthed in the United States. Although strands of the PG movement have existed for hundreds of years, its modern founder could be considered Oral Roberts.[2] Soon after World War II, Roberts began to encourage Christians to celebrate Christ's suffering on their behalf by being healthy, wealthy, and wise. Material wealth, he argued, could be a sign of God's favor, and believers should expect a miracle. This type of thinking is now considered mainstream in the Pentecostal faith, which is estimated to have as many as 250 million members worldwide, though not all such churches preach the prosperity gospel. What first took root among the poorer elements of the U.S. evangelical community now boasts believers from all walks of life and all nations. The rapid spread of Pentecostal churches in Africa, Latin America, and Asia is due in no small part to the attractiveness of the prosperity gospel, offering its good news to poor Christians around the globe in search of their version of the American Dream.[3]

The message behind the PG movement is simple: God wants Christians to prosper in all facets of their lives—including finances. Believers are encouraged to tithe (give 10 percent of their "first fruits") and make other financial donations to their church with the fervent hope that God will reward them with financial returns far above and beyond their original contribution. Oral Roberts called his followers to "seed faith," a term based on words Jesus spoke in the Gospel of Mark (10:30), where he said that those who gave up their possessions and followed God would "receive a hundredfold now in this time . . . and in the age to come, eternal life." As you can well imagine, there is much debate by theologians, pastors, congregants, and faiths as to the scriptural underpinnings of the prosperity gospel.

The lineage of this movement is much debated. Religious scholar Milmon Harrison sees its seeds in the Word of Faith (WOF) movement, which also emerged after World War II. The teachings of Kenneth Hagin Sr., Harrison asserts, make up the core beliefs of the WOF movement:

- The scriptures are in effect a contract between man and god. Believers are supposed to "name" what they want (health, wealth, wisdom, etc.) and "claim" the desired object by faith.

- The act of "naming it and claiming it" is an exercise in positive confession. Believers must be unshaken in their optimism regarding their desires and life circumstances.

- "Naming it and claiming it" in tandem with unbridled optimism leads to prosperity—health and material wealth. This is where the phrase "prosperity gospel" got its beginning.[4]

Twenty years later, Frederick Eikerenkoetter, better known as New York's "Reverend Ike," was spreading the gospel of success over the airwaves of 1,500 television and radio stations across America. Many successful evangelists struggle with money, generating boatloads of it but attempting to keep a low profile and at least the appearance of humility. Not Reverend Ike. The flamboyant preacher owned homes in New York and Hollywood, two mink-appointed Rolls Royces, two Mercedeses, a Bentley, heaps of jewelry, and a wardrobe that would have made Liberace blush—all financed by faithful givers. And you couldn't expect the good reverend to apologize. Conspicuous consumption, he argued, is next to godliness.

Ike's "Blessing Plan" was pure genius. Congregants were instructed to give whatever they could afford to Reverend Ike; and he assured them that it would be returned, with interest, to those of sufficient faith. The faithful were cajoled to pledge at least $100, even if they had to borrow it. They borrowed money for doctors and lawyers, he pointed out—so why not God? Send an extra offering and believe in extra blessings, Ike exhorted. To make sure his congregants received the blessings they so richly deserved, he allowed for blessings on the installment plan (weekly or monthly).[5]

The Blessing Plan allowed Reverend Ike to purchase a $600,000 theater in 1969 and spend $1 million renovating it, producing the newly named United Palace theater, with seating for 5,000. The Palace was used by Reverend Ike as a place of worship thereafter. Notwithstanding his flamboyant lifestyle, lavish wardrobe, and theatrical preaching, Reverend Ike might be best remembered for a long-running newspaper ad that touted his show *The Joy of Living*. The ad went like this: "If you want 'pie-in-the-sky when you die,' then Reverend Ike is not your man. If you want your pie now, with ice cream on top, then see and hear Reverend Ike on TV." I couldn't have summed up the prosperity gospel any better than that. Reverend Ike, recently deceased, saw his ministry hit its apex in the 1970s.[6]

The 1980s saw the meteoric emergence and equally high-profile fall from grace of the next generation of televangelists, with the likes of Jim and Tammy Faye Bakker and Jimmy Swaggart. Both the Bakkers and Swaggart were caught with their hands in the proverbial cookie jar. Jim and Tammy Faye, with Trinity Broadcasting Network founders Paul and Jan Crouch, created the *Praise the Lord* show for the Crouches' new Christian television network. The relationship was a rocky one, and the Bakkers left to begin their own show, *The PTL Club*. The show experienced rapid growth and was broadcasted by nearly 100 TV stations with audiences in the tens of millions.

By the early 1980s the Bakkers conceived and built the Heritage USA theme park in Fort Mill, South Carolina, and added a satellite system that enabled them to broadcast *The PTL Club* twenty-four hours a day across the United States. And the money flowed—weekly contributions estimated to exceed $1 million. Much of which were poured into the Christian-theme park and the ministries of *The PTL Club*. The theme park, however, would contribute to the Bakkers' fall from grace: from 1984 to 1987, Bakker and his colleagues sold $1,000 lifetime memberships to a luxury hotel on the park's grounds, entitling buyers to a three-night stay each year for their lifetime. Bakker and his associates were apparently very successful selling these exclusive memberships—so successful, in fact, that they sold more than there was space in the hotel. A large share of the revenue from the tens of thousands of memberships sold went into the operating budget of Heritage USA. Bakker did however keep "bonuses" totaling $3.4 million for his efforts.[7]

There's an interesting epilogue to the Bakker fiasco. In his 1996 book, *I Was Wrong*, Bakker renounced his past teachings on the prosperity gospel. He admitted that he had never actually read the Bible all the way through before going to prison and that he had taken key scriptural passages out of context in his message of prosperity. That message, Bakker mused, "did not line up with the tenor of scripture." He also expressed his gratitude to God for striking him down as a "false prophet."[8]

The names may have changed in this new century—you won't hear about the likes of Kenneth Hagin Sr., Oral Roberts, or Jim and Tammy Faye Bakker—but preachers of the prosperity gospel are by no means a fringe element of the modern religious scene. It's likely that you have seen or heard them on the Internet, the radio, or television; or you might be a member of a "megachurch" led by the likes of Kenneth Copeland, Benny Hinn, T. D. Jakes, Joyce Meyer, Creflo Dollar (and wife Taffi Dollar), or the increasingly popular Joel Osteen.

Many consider Kenneth Copeland the "godfather" of the prosperity gospel. He is an author, speaker, televangelist, and ardent proponent of the Word of Faith movement, having studied under its founder, Kenneth Hagin Sr. Copeland also founded and heads Kenneth Copeland Ministries, headquartered in Newark, Texas. He has borne the brunt of criticism for his PG teachings on wealth and his claim that true believers will receive a huge return on their tithes and donations to the church.

At a minimum, the PG movement has served Copeland and his wife well. They live in a lavish 18,000-square-foot lakefront home, have use of a God-ordained fleet of private planes and a ministry-owned airport, own a 1,500-acre ranch, and appear to lack none of the creature comforts that many of their "partners" (donors) in Christ no doubt go without. Copeland states, in regard to his newest jet, "The Lord spoke to me and said 'you're gonna believe for a Citation 10 [jet] right now.'" Copeland promises a hundredfold return in happiness and wealth for every kind donation. Indeed, the giving envelope states, "I am sowing $_____ and believing for a hundredfold return."[9]

Joel Osteen is the senior pastor of the 40,000-plus-member Lakewood Church of Houston, Texas. Osteen and wife Victoria's nondenominational Christian ministry reaches over 2 million viewers each week and is seen in over 100 nations around the world. Lakewood's home is the giant Compaq Center—formerly home to the NBA's Houston Rockets. After a $95 million (that's not a typo) renovation, it is now the headquarters of his approximately $60 million a year empire. Osteen is so popular that he was featured by *ABC News* as one of "10 most fascinating people of 2006"; he was also named "the most influential Christian in America" in 2006 by *The Church Report*. His books regularly top the *New York Times* "Best Sellers" list. His second book, *Become a Better You*, was released in 2007 with an initial printing of 3 million copies.[10]

Osteen's father, John, founded and developed Lakewood Church into a thriving body of approximately 6,000 members. The senior Osteen's ministry included a television broadcast, crusades, food pantries and other services for the needy as well as numerous other outreach programs. The younger Osteen earned a college degree in television production. Some point to his lack of any type of formal theological training. Dubbed the "Smiling Preacher," he offers sermons that take an optimistic approach to life with a touch of humor and with few mentions of sin or sacrifice—an approach that some call "theology light."[11] In the pantheon of today's PG preachers, I would rate Osteen a moderate, lacking the directness of his contemporaries, including Copeland, Dollar, Jakes, Meyer, and Hinn.

Bishop T. D. Jakes, another PG proponent, presides over a far-flung global ministry that includes his 17,000-member Potter's House Church in Dallas, Texas. The Potter's Touch carries his sermons around the world and runs on the Trinity and Black Entertainment Networks. Jakes's annual revival, known as MegaFest, draws 100,000 people. Jakes teaches that God rewards his faithful followers with material possessions. He is proud, and unapologetic, of the material possessions he has amassed—expensive suits and jewelry, fancy autos, homes, and land holdings. He has also written over thirty books, several of which have been best sellers.[12]

Creflo Augustus Dollar Jr.—you have to appreciate the irony of a PG teacher with the last name of Dollar!—is a thriving televangelist, pastor, and founder of the nondenominational Christian World Changers Ministries (WCM) located in College Park, Georgia. WCM, the parent organization for Dollar's 15,000-member World Changers Church International, has an annual budget of around $80 million as well as a myriad of other suborganizations, including Arrow Records. Dollar's church in College Park, which seats 8,500, is called the World Dome. It's purported that the $18 million church was built without any outside financing. And no one can claim that Pastor Dollar doesn't practice what he preaches. He owns a staggering number of big-ticket items, including a multi-million-dollar home in Atlanta, a $2.5 million apartment in Manhattan, two Rolls Royces, and a private jet. Dollar's teachings in regard to the PG movement have been coined the "gospel of bling."[13]

The prosperity gospel certainly isn't limited to these "big names" or to the Pentecostal faith. In fact, I recently attended a Catholic church where, at the end of the service, worshippers were given a small novena card that raised for me a PG warning. While a novena is normally a nine-day prayer, the term is sometimes used for any prayer that is repeated over a series of days. Given that my visit was during Christmas, I received the Saint Andrew Christmas novena card. On the front was a beautiful depiction of the baby Jesus, Joseph, Mary, and an angel. On the reverse was the Christmas novena, a short statement affirming the petitioner's belief that he or she would obtain what was asked as a result of the novena, and the following instructions: in order to fulfill the purpose of the prayer (stated as "to obtain favors"), the petitioner was to repeat the novena fifteen times each day until Christmas, asking God "to hear my prayers and grant my desires"; the petitioner's specific request(s) were to be included near the end of the prayer. The wording of this novena card suggests that even mainline religions have an element of the prosperity gospel in their teachings and rituals.

THE GOSPEL OF ASSIMILATION

Although the modern version of the prosperity gospel had its beginnings in the twentieth century, Christians have struggled with their relationship to money since the first century. As a result of what scholar Robert Franklin calls the "gospel of assimilation," churches have tended to adopt the values of the larger cultural environment and thus have drifted away from their original mission of service to others. This "mission drift" has no doubt been accelerated by the teachings of the prosperity gospel. Prominent pastors at many of America's largest churches sanctify the pursuit and accumulation of material possessions. In fact, most claim that God wants believers to be successful and have nice things. The core secular values of personal greed, materialism, and self-interest are being assimilated into the church. A world that not only permits but encourages vast disparities of wealth is now part of the gospel of many of our largest places of worship. "If preachers are preoccupied with pursuing the life of conspicuous consumption and preaching a 'prosperity gospel,' then poor people are in big trouble," quips Franklin.[14]

IS GOD TO BLAME FOR THE SUBPRIME REAL ESTATE FIASCO? Jonathan Walton, professor of religion at the University of California–Riverside, believes that the prosperity gospel (PG) may be partially to blame for the current mortgage crisis. Worshippers schooled in that gospel were the proverbial Christians led to the lions (unscrupulous mortgage brokers). Their belief that God would provide the resources necessary and remove any obstacles for them to enjoy the "better things in life" made them easy targets for banks and mortgage companies bent on expanding their own portfolios. A strong faith and regular donations to church were all that was needed for God to open the doors to a new home, they had been told.

!! CONSUMERS !!
!! GONE WILD !!

Clearly, many worshippers reasoned, it was divine intervention that had caused the banks and mortgage companies to overlook their faltering credit scores, bad debt, previous bankruptcies, foreclosures, or insufficient income, when in reality it was nothing more than bad (and greedy) banking practices. Readily available credit over the past ten-plus years appeared to validate the prosperity gospel's teachings. Finally it was time for the poor to ride the wave of economic prosperity taking place—or so it seemed to PG believers. But the banks, like all houses of cards, fell when the wind blew, leaving many worse off for their faith.[15]

Despite the post–World War II beginnings of the modern PG movement, it would be a mistake to view the prosperity gospel as a recent phenomenon. As religion scholar Ginger Stickney notes, "One can trace the roots of prosperity gospel to the late nineteenth century."[16] She points to the teachings of Henry Ward Beecher as evidence of early roots of the PG movement. Beecher was a hugely popular Congregationalist clergyman, social reformer, abolitionist, and itinerant speaker in the mid to late 1800s.

In his writings and sermons Beecher espoused a theology of what he called "economic Arminianism" (Arminianism being a Christian doctrine affirming each individual's free will). His theology's core tenet was that anyone who hoped for success (and salvation) had the wherewithal to achieve it. That individualism is likewise a foundational concept in today's prosperity gospel, and it taints how we see the poor and disadvantaged among us. Charity, in Beecher's thinking, is inherently dangerous, in that it discourages people from "going out and getting a job" (my words).

But aren't those two different things—religion and the economy? Yes and no. Religious scholar George Thomas's theory of isomorphism is helpful in understanding how the lines between the secular and nonsecular worlds end up blurred. Thomas argues that institutions respond to changes in the larger cultural environment and that different institutions may evolve analogously. The nineteenth century saw a transformation in this country from a largely agriculturally based economy to a manufacturing economy during the Industrial Revolution of the late 1800s. As capitalism seeped further and further into American life, the church "began to develop theologies that made sense in the broader cultural world."[17]

Preaching and writing during this pivotal period in American history, Beecher promoted a theology that reminded people of their "individual responsibility," the phrase with which he titled an important sermon delivered in 1870. In that sermon he talked about how God ordains wealth, the power of the individual, and that poverty is a sin. In speaking about individualism in a way that linked faith and the everyday world, Beecher laid the groundwork for the PG perspective that God gives each individual the wherewithal to succeed and that the individual is solely responsible for his or her own fate. Beecher exhorted that "humans are not bound by anything except their own limited notions of themselves."[18] This is strikingly similar to the teachings of leading PG teachers of today.

Beecher's ideas on work and individual responsibility in his "individual responsibility" sermon led to the notion adopted by contemporary prosperity gospel preachers that the accumulation of wealth is ordained by God. Beecher argued that God is not limited to the religious arena. He inspires and blesses our activities in the secular world as well. Russell Conwell, considered a cofounder with Beecher of economic Arminianism, was a staunch supporter of the idea that God sanctions wealth, claiming in a 1915 sermon that "to make money honestly is to preach the gospel."[19] He argued in that same sermon that it is our Christian and godly duty to have money. As Ginger Stickney summarizes the divine/human linkage, "God provides the inspiration while humans provide the gumption."[20]

It's clear from the teachings of Beecher, Conwell, and others not discussed here that the roots of the prosperity gospel are deeply ingrained in the American culture. A perversion of the spirit of economic Arminianism that characterized Beecher's teaching and the ethic of an industrializing nation in the late nineteenth century can be found in twenty-first-century churches where the prosperity gospel is preached. In the more extreme cases, the hard work component of Arminianism has been replaced with a focus on divine intervention: believers are asked to support the church with their tithes and offerings and then told to expect God's financial blessings. God is seen as a celestial ATM of sorts.

"MEGASIZE ME"

Ginger Stickney, in a book chapter titled "Godly Riches," writes that "religion . . . is a dynamic expression of society as a whole."[21] Nowhere is this statement truer than in the recent explosion in the growth of so-called megachurches. To be classified as a megachurch, a church must attract a minimum of 2,000 people to its weekly services. Twenty-five years ago there were only seventy-four churches in the U.S that attracted that number. Presently, there are more than 1,200 with an average weekly attendance of 3,500 worshippers. Megachurches have easily outpaced our nation's growth. According to researchers Alan Greenblatt and Tracie Powell, "Since 1900, the nation's population quadrupled while the number of megachurches increased 121-fold. In 2005 there were four megachurches for every million people—30 times more than in 1900."[22]

Pastor Robert Schuller's Crystal Cathedral, dedicated in 1980 in Garden Grove, California, signaled what many consider the birth of the *modern* megachurch juggernaut. His ongoing *Hour of Power* television ministry reaches millions of people across the globe. Emphasis on *modern* is key, given that pastors such as Charles Spurgeon and Henry Ward Beecher drew huge crowds with their fiery sermons and charismatic personalities.

A majority of today's megachurches are located on large suburban campuses, are easily accessed by an interstate or other major highway, have huge sanctuaries, and boast the latest in technology. Giant TV screens and sophisticated sound systems are de rigueur. Other amenities include gymnasiums, fitness and day-care centers, youth programs, and full-service cafeterias—sometimes even a Starbucks or Subway outlet. Most megachurches are located in the southern and western United States. California, Texas, and Georgia have the largest number of people worshipping at megachurches.[23]

Because of their use of multiple media channels—radio, television, the Internet, books—to spread the word, many megachurch pastors are household names. Indeed, with secular media coverage of their traveling crusades and their comings and goings, many megachurch pastors have reached rock-star status. As can be seen below, Houston's Lakewood Church, headed by Joel and Victoria Osteen, is the largest megachurch in the United States, with weekly attendance in the area of 40,000. No doubt you have caught one of Pastor Osteen's televised sermons. His books have sold millions of copies, his crusades are hugely popular, and he and his wife are media regulars.

The Largest Megachurches in the United States

CHURCH	PASTOR	LOCATION	AVERAGE WEEKLY ATTENDANCE	DENOMINATION
Lakewood Church	Joel Osteen	Houston, TX	43,500	Nondenominational
LifeChurch	Graig Groeschel	Edmond, OK	26,776	Evangelical Covenant
Willow Creek Community Church	Bill Hybels	South Barrington, IL	23,400	Nondenominational
North Point Community Church	Andy Stanley	Alpharetta, GA	23,377	Nondenominational
Second Baptist Church	Edwin Young	Houston, TX	22,723	Southern Baptist
Saddleback Valley Community Church	Rick Warren	Lake Forest, CA	22,418	Southern Baptist
West Angeles Church of God in Christ	Charles Blake	Los Angeles, CA	20,000	Church of God in Christ
Fellowship Church	Ed Young, Jr.	Grapevine, TX	18,355	Southern Baptist

CHURCH	PASTOR	LOCATION	AVERAGE WEEKLY ATTENDANCE	DENOMINATION
Southeast Christian Church	Dave Stone	Louisville, KY	17,261	Christian
Fellowship of the Woodlands	Kerry Shook	The Woodlands, TX	17,142	Nondenominational
The Potter's House Church	T. D. Jakes	Dallas, TX	17,000	Nondenominational
Phoenix First Assembly of God	Tommy Barnett	Phoenix, AZ	16,000	The Assemblies of God
Hopewell Missionary Baptist	William Sheals	Norcross, GA	16,000	Baptist
Calvary Chapel	Robert Coy	Fort Lauderdale, FL	15,921	Calvary Chapel
Central Christian Church	Jud Wilhite	Henderson, NV	15,081	Christian
First Baptist Church	Jack Schaap	Hammond, IN	15,059	Baptist
Harvest Christian Fellowship	Greg Laurie	Riverside, CA	15,000	Calvary Chapel
World Changers Church International	Creflo Dollar	College Park, GA	15,000	Nondenominational
Prestonwood Baptist Church	Jack Graham	Plano, TX	14,975	Southern Baptist
New Light Christian Center	I. V. Hillard	Houston, TX	13,500	Nondenominational

Source: Hartford Institute for Religion Research, 2010.

All of the pastors listed above are engaging and charismatic, and many (though not all) preach the prosperity gospel—more a message of personal fulfillment and success than theology. One high-profile exception to megachurch leaders proclaiming the prosperity gospel is Pastor Rick Warren of Saddleback Valley Community Church. Author of the book *The Purpose Driven Life: What on Earth Am I Here For?* which sold 25 million copies, Warren finds the premise of the prosperity gospel laughable. He calls the idea that God wants everyone to be wealthy "baloney." Warren claims, "It's creating a false idol," and asks, "[W]hy isn't everyone in church a millionaire?"[24] Critics voice concerns that many megachurches have eschewed the more traditional religious mission of worshipping God and serving others and are not only extolling (in both word and deed) the virtues of worldly goods and wealth but are focusing on helping worshippers get rich. As one critic observed, "In this country, the Christian faith has succumbed to consumerism."[25] As the rise of capitalistic and materialistic ideals continues (the current recession a mere bump in the road) into the new century, so does a theology to complement it.

And it's a message that people want to hear. A 2007 survey of Protestant churchgoers conducted by the Institute for Studies of Religion at Baylor University in Waco, Texas, found that 57 percent of megachurch members agreed with the statement "God rewards the faithful with major successes." Now, we could argue until the cows come home as to the meaning of "major successes," but I feel comfortable that most of us would include money as an integral component of that definition.[26]

A large survey published by *Time* magazine in 2006 as part of an article on the prosperity gospel found similar results. A solid 61 percent of Christians surveyed agreed with the statement "God wants people to be financially prosperous," while 31 percent agreed with the statement "If you give away your money to God, God will bless you with more"—a key promise of the prosperity gospel. On the other hand, 49 percent disagreed with the statement "Poverty can be a blessing from God" and 44 percent disagreed that "Jesus was not rich, and we should follow his example."[27]

Like all good retailers, megachurch pastors must be schooled in the ways of marketing. In addition to the various communications media needed to spread their message beyond church walls, their product offerings must also hit the target. The worship service itself is typically more like a rock concert than anything else. Not surprisingly, then, insiders say that the real heart of any megachurch service is its technical staff. Nothing is left to chance—the sound, lighting, and camera angles must all be just right. This takes the skills of seasoned, well-paid event-staging professionals. Not to worry: *Technologies for Worship Magazine* (*TFWM*), with a circulation of 25,000, sponsors an annual conference, with sessions covering such topics as:

- Web-casting production for houses of worship

- Basic acoustics for houses of worship

- Introducing change without offending *almost* anyone

- Wireless microphones for houses of worship

- Pastoring the tech team

- Enhancing the experience: developing immersive atmospheres using video

- Show coordination and show production

- Budget-stretching solutions

The use of the word "show" in referring to worship services (as in the next-to-last list item) always makes me uncomfortable, as does the phrase "budget-stretching solutions" (final item). Apparently a lot of people are fine with such lingo, though. National Media Finance's ad in the 2009 *TFWM* conference program "offers budget stretching solutions to help churches spread equipment costs to match the timing of tithing, contributions, and donations." Saving souls is a serious business.

In addition to a seamless show, megachurches have a message—but it's not necessarily what you'd expect. Pastor Lee McFarland of Surprise, Arizona, oversees a flock of 5,000 worshippers and preaches less about theology and more about how to overcome the challenges of every-day life. His sermons touch on such subjects as disciplining children (where do I sign up?), setting and reaching life goals, reducing debt, and even breaking addictions to pornography and other bad habits. McFarland says, "If Oprah and Dr. Phil are doing it, why shouldn't we?"[28]

James B. Twitchell, a professor of English at the University of Florida and a social critic, likens what megachurches have done to "what Walmart did to merchandising," adding that they have become "the low-cost deliverer of salvation."[29] Clearly, megachurches are meeting the needs of the consumer—uh, make that worshipper—the essence of any successful marketing campaign. Their size allows them to capitalize on economies of scale, meaning that they can do things that smaller churches don't have the resources to do. A typical church with one to two hundred worshippers simply can't afford to offer the panoply of programs, music, and entertainment that are characteristic of megachurches. And the people have spoken: while overall church attendance has remained steady, megachurches have grown at the expense of smaller churches. It seems to be what's known as a zero-sum game—in other words, any growth in megachurch attendance comes at the expense of its competitors, smaller churches. Sounds like the battle between big-box retailers such as Walmart and local mom-and-pop businesses, doesn't it?

The blurring between the secular and nonsecular worlds is nowhere more evident than in megachurches. Megachurch members are a coveted target market for "Christian-oriented" products as well as more mainline secular offerings. Marketing campaigns for Christian-themed movies are often kicked off in megachurches. The movie *Evan Almighty*, starring Steve Carell from *The Office* sitcom fame, dedicated a good portion of its marketing budget to the megachurch market. Mel Gibson's *The Passion of Christ* did $370 million at the box office in the United States with a lot of help from the megachurch community. Rick Warren, founder of Saddleback Church in California, "hosted a screening for over 4,500 pastors and separately purchased 17,000 tickets for its congregants." Gibson's use of megachurches to market his film is now the marketing model used by others. Conversely, megachurches are using popular movies to carry the Christian message. Stories of megachurches purchasing large numbers of tickets to the local cinema or renting movie screens to watch Christian-themed movies inside the church are common.[30]

The marketing onslaught in megachurches does not end with Christian-themed products and services, however. Megachurches "have become a prime distribution and marketing channel, with companies including McDonald's, Ford, Chrysler, Target, and Coca-Cola, lining up to sponsor conferences and outreach efforts, as well as offering free samples and test drives directly to worshipers," according to Alan Greenblatt and Tracie Powell.[31] With access to mass markets becoming increasingly difficult, being able to communicate with as much as two-thirds of the U.S. population through churches has marketers salivating. Marketers appreciate large, easily accessible, and like-minded gatherings of people—just the kind they can find in the network of 1,200 megachurches strewn across the U.S. landscape.

T. D. Jakes, bishop of the 17,000-member The Potter's House Church in Dallas, Texas (the eleventh-largest megachurch in the United States), has garnered corporate financial support from such *Fortune* 500 companies as Ford, Coke, American Airlines, and even Pine-Sol to sponsor his hugely popular MegaFest religious festival, an annual event that draws more than 100,000 people during its run. Past festivals have included the likes of decidedly secular types including basketball great Magic Johnson, financial guru Suze Orman, and even singer Patti Labelle. Before its recent financial problems, Chrysler offered test drives of its new offerings at four of the nation's largest African American megachurches. Parishioners who test-drove a vehicle received free tickets to a Chrysler-sponsored gospel tour, and a $5 donation was made to a cancer center. Sounds like we've added retail commerce to the plethora of services already offered by megachurches. Talk about "mission drift"! "With 330,000 churches in America, it's potentially the largest distribution network in the country," says A. Larry Ross, who heads up a Dallas, Texas, marketing company which helps its clients sell their wares to the Christian market.[32]

Barry Harvey, a professor of contemporary theology at Baylor University, in Waco, Texas, had this to say in the *CQ Researcher* about the marriage of the secular and the nonsecular that's occurring in megachurches across the country: "It's blurring lines, the church is essentially becoming indistinguishable from its biggest competitor, the mall. To allow the commercial enterprise to come into the church is to allow the desire for accumulating things, buying things, to dominate even the relationship with God."[33]

I'll go you one better: we want megachurches, claims Lee McFarland (the previously quoted pastor who invoked Oprah and Dr. Phil), "to look like a mall. . . . We want you to come in here and say, 'Dude, where's the cinema?'"[34] Driving to a one-stop place of worship like a megachurch is akin to driving to the mall—something Americans are accustomed to doing. Critics of this secular/nonsecular marriage see it as "the latest lurch in Protestantism's on-going descent into full-blown American materialism."[35] Edith Blumhofer, director of the Center for the Study of American Evangelicals at Wheaton College, adds that "you don't have to give up the American Dream. You just see it as a sign of God's blessing."[36]

The consumer culture, the prosperity gospel, and modern megachurches enjoy a symbiotic relationship. The megachurch has adopted the prosperity gospel as a means of attracting the materialistic denizens of the current consumer culture. As the differences between the secular and nonsecular worlds continue to blur, it becomes hard to distinguish one from the other. We are being increasingly left with one worldview—that the pursuit, acquisition, and consumption of material possessions will serve us well both in this world and beyond.

Nothing the megachurches have done to increase their membership rolls is new. The use of market segmentation, careful research, attractive settings with lots of amenities, high-production-value services, and giving the customer what he or she wants is nothing new. These are tactics that marketers have used for years.

Now, in the next chapter, we'll look at how three particularly effective—and, some may argue, ethically questionable—strategies are used every day to further fan the flames of consumption.

WEAPONS OF MASS CONSUMPTION

If Moses had been handed the Ten Commandments on floppy disks, would word from on high have survived through the ages?

—EDWARD C. BAIG

Gerri Hirshey's story is one we all can relate to. It was time for her to buy a new stove—again. Her old stove of twelve years had finally given up the ghost, prompting a visit from the local fire department when it gasped its last breath and filled Hirshey's kitchen with gas from a broken burner. After the firefighters had given the all-clear to return to her kitchen, she summoned a repair technician (at $90 just to step in the door) and braced herself for an expensive repair bill.

The news was worse than she'd expected. After only a cursory inspection of the offending appliance, the repair tech suggested a mercy killing. Hirshey was informed that getting under the "hood" of her stove would require drilling and as many as twelve new screws. Not too bad, right—I mean, how much can screws cost? Well, all this would add up to approximately $600 in parts and labor, plus the cost of the initial house call. In total about $700 to repair the stove—more than its original cost. This stove, like so many other products in use today, wasn't built to be repaired—simply to be used and discarded once its useful life was over. Products are made with a predetermined life span—a "death date," if you will. We call this "planned obsolescence."[1]

Despite the urge to point our fingers at manufacturers for the shortened life span of many modern appliances, consumers are also partly to blame for the current state of affairs. We live in a "throwaway society" of our own making. This nation's shiny-objects ethos demands that our appliances and other products have the latest and greatest bells and whistles—a desire that industry has greedily fulfilled and Madison Avenue has aggressively marketed.

If we hope to mute those bells and whistles and stop the runaway train of consumerism, we've got to ignore all that marketing. But if we're going to resist Madison Avenue, the first thing we've got to figure out is what tricks marketers are going to use to get us to part with our hard-earned cash. In the words of the ancient Chinese military strategist Sun Tzu, we've got to know our enemy to defeat him.

We are deluged constantly with marketing messages—some subtle, some not. A wise man once said, "Advertisers constantly invent cures for which there is no disease." That may be a slight exaggeration, but certainly marketers are a sneaky lot. Remember: I *am* one. We have poked, prodded, and studied consumers for so long that we have a pretty good idea of what makes them tick. Time and space don't permit me to talk about all of marketing's "weapons of mass consumption," but I'll focus on three strategies we use to keep you spending: planned (or perceived) obsolescence, product placement, and spreading the word (word of mouth). These strategies and many others continue to prime the economic pump and keep the majority of us living blindly from paycheck to paycheck.

By the end of this chapter you will see that we marketers are indeed tricky and that knowledge of our wily ways is truly power. Armed with that knowledge, in chapters 12 through 14 we will take a look at how, with a modicum of self-control, you can outsmart the marketing machine and avoid the most common materialist pitfalls.

DESIGNED FOR THE DUMP

You know the story. Products that seem to break long before they should or go out of style seemingly overnight. But what you might not know is that the premature death of products is a tried-and-true business strategy. It's more than just a coincidence that so many products die just about the time their warranty expires. Or that replacing a single component costs more than the original purchase price of the product. Fix a toaster? It isn't worth it. And product life has gotten even shorter in the digital age. We all quickly moved from floppy disks to CDs to store important information. But what about CD rot? Will our treasured family photos stand the test of time? Computers are nearly obsolete before we take them out of the box; iPod batteries lose peak functionality in about a year. The list of products designed for the dump goes on forever. What about cell phones? Over 100 million of them are discarded annually in the United States alone; they either quit working or go out of style. Everyone wants to foster the proper image, and that image needs to be constantly updated.

We are, indeed, a nation of sheep. Computers, cell phones, iPods, and CDs are only a few of countless examples of how our behavior as consumers is dictated by the marketing strategy of perceived obsolescence. This isn't news to anybody, I'm sure. A good question would be, then: Knowing all this, why do we still follow the dictates of the marketplace?

When I teach on the subject of consumer behavior, I use something called the "consumer purchase decision process" (CPDP), shown below, to explain the steps consumers pass through when making purchases. The amount of time and effort spent as one progresses through the five-step CPDP varies greatly depending upon whether it's a purchase decision you make regularly ("programmed" decisions such as shopping for food and buying gas) or one you make only infrequently ("non-programmed" decisions such as buying a house, purchasing life insurance, or deciding on a nursing home for your aging parent).

Consumer Purchase Decision Process

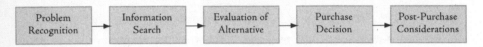

Whether you're making a routine purchase for the thousandth time or making a first-time buy of something complex, the CPDP begins with the problem recognition stage. At this initial stage, the consumer's current and ideal states are out of balance. For example, your son or daughter is leaving for college and needs to replace a terribly out-of-date laptop computer (maybe all of two years old). The tension created by the imbalance between current and ideal states triggers behavior. It's the job of marketers and advertisers to create such tension. Two techniques that we have found to be particularly effective are planned obsolescence and perceived obsolescence. Let's look at each technique in turn.

Planned Obsolescence

Planned obsolescence (PO) is a design strategy of intentionally developing and marketing products with a limited life span. The objective of planned obsolescence is to stimulate replacement purchases by consumers. The easiest and most straightforward way to speed up replacement purchasing is to shorten the usable life of a product through one or more of the following three strategies focused on the functioning of the product itself:[2]

1. DESIGN THE PRODUCT TO FAIL WITHIN A CERTAIN PERIOD OF TIME ("DEATH DATING"). The average life span of a cell phone is about eighteen months. And your iPod? Apple openly admits that it's designed to last only three years—and that's if all goes well. In fact, the lithium batteries used in iPods begin to lose peak functioning after a year of use. What about your toaster, dishwasher, refrigerator, or hot-water heater? Since we've bonded in the course of many chapters, I'll open up and tell you about my experience with what I call "blow-combs"—hand-held hair dryers—back in the 1960s and 1970s. I can't tell you how many blow-combs I went through! And it was always the same story: the blow-comb would fail at some crucial moment in a burst of smoke and sparks. As a compliant consumer, I would dutifully throw the offending item into the trash and buy another.

But I'm just getting started. During the useful life of the blow-comb itself, its attachments (various combs and brushes) would routinely break, leaving me with two undesirable alternatives. I could attempt to find replacement attachments and pay a good share of the original purchase price for what should be a 25¢ plastic gadget. Alternatively, I could simply chuck it all and buy a new blow-comb set complete with all the attachments—PO at its finest. I can't tell you how many times I used an ill-fitting comb from an earlier purchase just so those darn marketers wouldn't get the best of me.

2. MAKE REPAIRS DIFFICULT AND EXPENSIVE.   When's the last time you had anything repaired? Despite the fact that we purchase approximately 34.5 million new TVs each year, only 20,000 are repaired. We are forced to dispose of literally mountains of notebook computers, digital cameras, iPods, cell phones, and DVD players because repairs are too expensive (if we could even find someone to work on them). A shocking statistic about our throwaway society comes from international sustainability expert Annie Leonard's animated documentary *The Story of Stuff* (www.storyofstuff.com), which examines our consumer society. Leonard discovered that after six months, only 1 percent (!) of all products we purchase is still in use. A full 99 percent has been relegated to dumps around the world—or worse, is polluting our natural environment. Did you know that there's a mass of plastic roughly the size of Texas floating in the ocean as a result of our love (and discarding) of plastic water bottles, plastic bags, and the like?

Our abandoned cell phones, computers, and other electronic gadgets—those that we don't simply toss in our household trash—are shipped to countries like China, India, and Bangladesh, where low-paid workers (often women and children) are hired to salvage reusable parts. These giant mounds of electronics contain mercury, lead, flame retardants, and PVC-coated wire, and the extraction of usable parts releases these toxins into the groundwater and the air (and thus into the lungs and skin of workers). What can't be salvaged is burned or dumped, both of which pose significant environmental risks.

3. DESIGN THE PRODUCT IN A WAY THAT LEADS TO REDUCED SATISFACTION.

Research tells us that an unappealing product can spur premature disposal. This is a particularly ingenious ploy—one I had noticed but hadn't really thought of as an intentional design feature until I began my research on PO. I'd noticed, for example, how quickly new cars with perfectly pristine, smooth, and highly polished surfaces became damaged with everyday use, prompting dissatisfied customers and premature replacement. Nothing looks better on a car than a large expanse of highly polished metal, until even the slightest ding interrupts its flow. The same holds for household appliances—a major component of the world's landfills. Let's face it: those pristine surfaces are not going to last long with everyday use—or worse, with the hard use that children can impose.

Perceived Obsolescence

While the planned obsolescence strategies above focus on sabotaging the product itself—the first half of a lethal one-two combination—strategies of perceived obsolescence focus on encouraging dissatisfaction with current products:[3]

1. DESIGN FOR FASHION.

Although men's clothing styles come and go and I'm able to cope, my head begins to spin and I get dizzy when thinking about women's fashions. We have already discussed fashion styles from the 1970s, and we could have the same discussion about any ten-year period since the Industrial Revolution—that's how dependable changes in fashion are. Clearly, outside of the need to buy a few new undergarments, shoes, pants, and tops occasionally, most spending on clothes is discretionary. It is the two-pronged job of marketing to both extol the wondrous properties of the newest apparel and assail our present fashion statements. As noted earlier, appealing to people's insecurities is the second-most common advertising approach (after price). Advertisers use the "bandwagon" approach: simply put, everyone else is doing it—why not you?

But don't for a moment think that designing for fashion ends with clothes. What about cars? General Motors is considered the inventor of perceived obsolescence. By the 1920s it had adopted regular changes in both technology and style, as we saw earlier; and by 1927 GM president Alfred Sloan had wholeheartedly embraced the yearly style change, later referred to as "artificial obsolescence." At the time, Charles Kettering, the head of GM's Research Division, defined that division's mission as the "organized creation of dissatisfaction." General Motors overtook Ford in sales because of its focus on yearly model changes and Henry Ford's insistence on positioning his automobiles as durable and economical.[4]

A slowdown in the economy during the early 1950s meant that a stimulus was needed to spur consumption. That stimulus came when artificial obsolescence expanded beyond the automobile industry to consumer goods in general. A *Business Week* article of 1955 heralded this development, stating that "planned [and presumably perceived] obsolescence is here to stay in the auto industry and it is moving into more and more fields."[5] The 1950s saw the birth of the throwaway ethic that has grown exponentially to this day. The industrial design profession was called on to create new fashions or styles for all sorts of products, as well as to design products for failure. The design profession played an important role in the development of the broadening consumer culture of the 1950s, as it continues to do today. Vance Packard's now classic book *The Waste Makers*, cited earlier, chides business for its systematic attempt to encourage a throwaway ethos in consumers and for promoting consumer debt and discontent.

It's hard to find any product category that isn't impacted by style changes. Presently, fashion drives an incredible turnover in consumer durables such as MP3 players, laptop computers, and of course cell phones. My daughter likes to festoon her phone with shiny objects, stickers, and other doodads to keep up with the current definition of "cool." Cell phones as fashion statements, notes designer Terry Irwin, are "the ultimate contradiction in design: ephemeral products made of 'permanent' materials."[6]

2. DESIGN THE PRODUCT FOR FUNCTIONAL ENHANCEMENT. Manufacturers often deliberately build in the need for enhancements or upgrades. We've heard this tune played so many times it's now a classic. Technology has shortened the time it takes to render old products obsolete—at least in the eyes of the consumer (and with a little help from marketers). Again, think of cell phones. Originally used simply to call people in case of emergency or when you were away from home or office, cell phones are now all-purpose devices. They allow you to take pictures, listen to music, keep a calendar, send text messages or e-mails, surf the Web, play games, and so on; Lord only knows what the next "breakthrough" use will be. From a marketing perspective it really doesn't matter. As long as that new use creates tension and longing in consumers and translates into replacement purchases, marketers have done their job. Research tells us that more rapid introduction of upgrades motivates faster replacement of old products, regardless of the level of quality enhancement. Nowhere is this tendency clearer than in the marketing of cell phones, but it's the same story for technology and computers: laptops are now faster, have more memory, are physically smaller and lighter, and come in an array of styles and colors. Clearly, the strategy of perceived obsolescence is working in that field, since North Americans throw away approximately 300 million PCs each year.[7]

The solution, if there is a viable one, to planned and perceived obsolescence is not readily apparent. Our economy is addicted to growth, and engineered obsolescence is at the very core of how we have to this point maintained such growth. The more reliable and durable a product is, the longer the repeat purchase cycle becomes—and thus the slower the firm's sales growth. "Addiction" is not too strong a term for our economy's need to expand. Designing for obsolescence (1) allows firms to stimulate sales growth through shorter replacement cycles, (2) reduces competition from the used-goods market (the products aren't around long enough to pose a threat), and (3) gives firms an excuse to increase prices for "improved" replacement products.

And this isn't just industry's problem. A large-scale survey of consumers by researcher Tim Cooper found that the life span of household appliances (refrigerators, dishwashers, microwaves, blenders, freezers, vacuums, etc.) is determined by consumer behavior as much as by how those products are made. Although we're beaten over the head by a constant barrage of marketing efforts designed to get us to spend, we consumers must take at least partial responsibility for the current mess. It's not easy to say no, but it can be done. We're all-too-willing participants in the "designed for the dump" culture. As Cooper concludes, "Planned obsolescence, though often disparaged, has long been tolerated."[8]

HIDING IN PLAIN SIGHT

Would you tune in to a radio show whose sole purpose was to sell Vaseline products? Probably not. But many people did, unbeknownst to them. In 1928, NBC created a radio show called *The Real Folks of Thompkins Corners* for the Chesebrough Manufacturing Company. NBC's goal with *Real Folks* was to spotlight Chesebrough's line of Vaseline products. With each episode the listening audience was allowed to eavesdrop on the "real" lives of Matt Thompkins—town mayor and gadabout—his lovely wife, Martha, and a cast of average small-town Americans who were never at a loss for new uses for Vaseline products. Indeed, the various uses depicted carried out the raison d'être for the entire show: sell Vaseline products. Matt used Vaseline petroleum jelly to shave his uncommonly smooth face, and Martha recommended that Elmer put some Vaseline on his ailing toes. NBC and Chesebrough both felt that "concealing" the selling message within the story line would be more effective and credible than touting a product blatantly in the traditional-format commercial. And they were right.[9]

"Product placement," a type of advertisement where products or services are inserted without fanfare into TV shows, movies, books, video games, and various other media, has long been a staple in the marketing arsenal. The recent explosive growth in the use of product placement, however, has many critics and consumer groups crying foul. The product placement industry is estimated by some to be as large as $10 billion in annual sales.[10] Nary a TV show, movie, or video game is made anymore without product placement—and generally a lot of it. It's just another way that marketers can make a buck in an increasingly competitive marketplace.

Because consumers may not be aware of the selling intent of such placement, they are highly susceptible to its message. The purported influence of product placement on consumers is so great that its detractors have lobbied for stricter federal regulation of the practice. Laws are already in place that require licensed broadcasters to disclose to their audience when they (or their staff) receive money or other goods for on-air promotion of a product. Section 317 of the federal Communications Act states that "all matter broadcast by any radio station for which money, service, or other valuable consideration is directly or indirectly paid, or promoted to, or charged or accepted by, the station for broadcasting, from any person, shall, at the time the same is so broadcast, be announced as paid for or furnished, as the case may be, by such person."[11] This law evolved from the so-called payola scandal of the late 1950s, in which radio disc jockey Alan Freed and others were accused of accepting money ("payola") to play certain songs. Freed was found guilty and fined, and the above section was enacted to stop such behavior in the future.

Great Britain has banned product placement. I see no such fate for product placement in the United States, where it is already so firmly entrenched. A U.S. ban has its supporters, however: various consumer groups have lobbied the Federal Communications Commission (FCC) to investigate the practice of product placement. In June of 2008, for example, twenty-three consumer groups sent a letter to the FCC asking it for tighter restrictions on what they refer to as "advertainment."[12]

Although the wheels of justice turn slowly, the FCC has begun to address the product placement controversy and is considering new sponsorship regulations. For instance, it is considering requiring advertisers to provide both visual and audio notices that a product's appearance in a show is a paid ad—and not at the beginning or the end of the program (or in small print that flies across the screen), but when the product appears. At the current rate of product placement in many shows (often multiple exposures of the same brand)—approximately 109 per episode of everybody's favorite, *American Idol*, and a staggering 250 per episode of *The Biggest Loser* (4,000 over the course of sixteen shows)—this may be an untenable proposition.[13]

The consumer activist group, Commercial Alert, has been haranguing the FCC for years, claiming that product placements are "dishonest advertising."[14] *Adbusters* magazine, an anti-advertising publication, feels that it is unethical to market products to Americans without them knowing it. Children are seen as especially susceptible to such practices for two reasons: (1) kids have not yet developed a working knowledge of this kind of promotional tool, and (2) as a rule, they are more susceptible to product placement than adults. The vulnerability of children arises because they lack the ability to tell the difference between show content and product placement.[15] Although many of us parents have criticized much of children's programming as nothing more than extended commercials, few of us realize what a significant effect it has on our children. Because children are completely unaware that they are being exposed to commercial messages, product placement impacts them at an almost subconscious level—which immediately brings to mind the late-twentieth-century controversy surrounding subliminal messages.

Subliminal messages are pictures—of a product, say—embedded ("hidden") in newspaper and magazine ads or run across a movie or television screen at such a high speed that they can't be consciously perceived. The intent, when this practice was first implemented, was that these messages would somehow be processed at a subconscious level, propelling us to blindly follow the advertiser's suggestion to drink more Coca-Cola or eat more Kentucky Fried Chicken. Alas, the effectiveness of subliminal messages and their use in advertising has largely been debunked.

Subliminal messages and product placements aren't quite the same, though. Whereas subliminal messages are supposed to affect consumers at a completely subconscious level, advertisers want consumers to notice—and, in the case of video games, to interact with—their products. The objective of good product placement, and there is a lot of bad placement as well, is to seamlessly weave the product into the story, cleverly disguising the selling intent. Blurring the line between advertising and entertainment is an ethical issue not just for children but for adults as well.

As I mentioned earlier, product placement is not a new phenomenon. The emergence of radio as a mass medium in the 1920s gave birth to this particular marketing strategy. With lots of time to fill, radio networks looked to advertising agencies and to network sponsors (advertisers) to create shows. And they delivered, producing programs designed to spotlight their products. In addition to learning all of the various uses for Vaseline on *The Real Folks of Thompkins Corners*, listeners to that show were tempted by the Log Cabin Company of syrup fame, who sponsored *Real Folks* in 1932. A person turning the dial in that era might have encountered such radio shows as Kraft Music Hall with host Bing Crosby, the Coca-Cola Top Notchers (1930-32) and the Pause that Refreshes (Coca-Cola 1934-35), the Palmolive Hour, Champion Spark Plug Hour, General Mills, Betty Crocker, who had her own show, and Aunt Jemima, whose adventures were recounted on a Quaker Oats–sponsored show set on the "old Higbee Plantation in Dixie," which was famous for "miles around because of Aunt Jemima's cooking." You can bet that Quaker Oats rarely missed an opportunity to fold its pancake mix into the lives of the residents of the Higbee Plantation.[16]

This single-sponsor advertising model quickly took hold, and by 1929 advertising agencies created and produced 55 percent of all programs on the radio. The daytime radio schedule was filled with popular serials, many referred to as "soap operas" because they were sponsored by companies like Procter & Gamble that sold soap products.[17] Given free rein in regard to a show's content, advertising agencies ran amok, writing the scripts, hiring the performers and producers and directors, and designing entire shows around the sponsor's product(s). Product placement within programs became a routine occurrence and advertisers used a heavy hand in censoring dialogue and manipulating story lines. The broadcast networks had largely ceded control of a show's content to its sponsors and their advertising agencies.

When the fledgling medium of television rose in the late 1940s and early '50s, radio was its obvious business and programming model. Many of radio's shows and stars made their way to the small screen, a solid use of advertising as a primary source of revenue. By the late 1950s, more than one-third of all television programs were created and controlled by advertisers and their ad agencies.[18]

The ceding of control to TV advertisers came at a price. Take, for example, Camel cigarettes' sponsorship of the television show *Man Against Crime*, starring Ralph Bellamy (before my time too). The show's writing, story line, characters, and stars were carefully orchestrated to place the sponsor's product in the most favorable light—a practice that would have dire consequences for the health of the nation. The following quote describes the instructions issued to the writers, directors, and actors of that show: "Cigarettes were never to be associated with 'bad' or disagreeable characters or plots[;] . . . there was to be no suggestion of a narcotic effect; arson and fires were never to be mentioned."[19]

Mr. Television himself, Milton Berle—as his nickname would suggest—came to prominence with the emergence of television as the new mass media. In the early 1950s Berle was the host and star of *Texaco Star Theatre*, which began each week with the following ditty:

Oh, we're the men of Texaco; we work from Maine to Mexico.
There's nothing like this Texaco of ours.
Our show tonight is powerful.
We'll wow you with an hour full of howls from a shower full
of stars.

A few years later, Mr. Television could be seen in the Buick automobile–sponsored *Buick-Berle Show*. Not straying far from the success of Berle's earlier show, *Buick-Berle* featured a lengthy opening sequence that associated the star with the car in no uncertain terms. To much fanfare the announcer would exclaim, "It's *The Buick Show*! It's *The Berle Show*! It's *The Buick-Berle Show*!" while the two names lit up the screen. Then, as a chorus belted out, "How I love to drive my Buick," Milton rode down Main Street USA in a Buick convertible.[20]

To see how inextricably intertwined entertainment and advertising had become we need look no further than the title of some other programs of the era, including *Maxwell House Hour, Palmolive Hour, Goodyear Television Playhouse, Kraft Television Theater, U.S. Steel Hour, Colgate Comedy Hour, Pepsi-Cola Playhouse*, and *Schlitz Playhouse of Stars*, to name but a few.

But television's close relationship to sponsorship quickly went beyond entertainment programming, as even "hard" news stories got in the game. The 1950s *Camel News Caravan* with John Cameron Swayze was a carefully orchestrated affair. The show opened with these words: "The makers of Camel cigarettes bring the world's latest news events right into your living room. Sit back, light up a Camel, and be a witness to the happenings that made history in the last 24 hours. Produced for Camel cigarettes by NBC."[21] In addition to the above opening, Camel ads were interspersed throughout the show, with a singer warbling "So mild, so mild" to be sure no one missed the undeniable benefits of Camel cigarettes. My favorite product placement was the lit cigarette that stayed in an ashtray on Swayze's desk throughout the show; during the show's signature closing, "John Cameron Swayze saying good night for Camel cigarettes," the camera would zoom in on that cigarette, just for good measure.

The good times and close relationship between entertainment and advertising were about to end, however. Television network executives grew wary of a system of sponsorship that put enormous power in the hands of advertisers, who not only expected creative control over sponsored shows but could withdraw their support with next to no notice. Pat Weaver, who became president of NBC in 1953, saw a better way.

To wrest power from advertisers and their ad agencies, Weaver increased the length of shows from the standard fifteen minutes used in radio to thirty and sixty minutes. With the cost of sponsoring an entire program now prohibitively high for most companies, he pushed for an advertising structure based on the print media model. In other words, rather than being the sole sponsor for a show, advertisers bought time within a program created by the television network itself. During his tenure at NBC, Weaver was responsible for creating the *Today* and *Tonight* shows, both of which allowed advertisers to pay for advertising space rather than sponsor an entire show.[22]

But it was the rise and fall of quiz shows in the late 1950s that ultimately put an end to single-sponsor programming. The 1957–58 television season featured approximately two dozen shows in this new quiz format, many of which were sponsored by single companies. Revlon's wildly popular *$64,000 Question*, for example, featured a set conspicuously adorned with the company's name. But both the quiz show format and the single-sponsor model that made it possible were brought down by one man, Charles Van Doren. His phenomenal success on another Revlon quiz show, *Twenty-One*, garnered a large following, but Van Doren later confessed that he had in fact been fed many questions and answers by the show's producers. Once the cat was out of the bag, producers admitted that Revlon had often dictated which contestants should continue and which should not. What followed was nothing less than a national scandal (immortalized in Robert Redford's docudrama *Quiz Show*), complete with Senate hearings. It also heralded the end of single-sponsored TV shows; by the late 1960s only a handful still existed. The golden era of TV sponsorship had ignominiously come to an end.[23] But the hiatus of using products in programming was brief, and that practice reemerged with such a vengeance that it would have appalled pioneers of the art.

The coming-out party for product placement in movies came in the form of Steven Spielberg's 1982 blockbuster *E.T.* The shy extraterrestrial was coaxed out of hiding with the aid of Reese's Pieces—a new candy introduced by Hershey's. The $1 million investment paid off handsomely for the candy maker: sales of the peanut butter goodies skyrocketed by 80 percent.[24] That success marked a shift in the film industry's profit model as other companies looked to replicate the success of Hershey's foray into product placement. Other noteworthy, and some say classic, examples of product placement include Ray Ban sunglasses in *Top Gun*, *Risky Business*, and *Men in Black*, FedEx in the Tom Hanks movie *Cast Away*, and the Jamaican beer Red Stripe in the John Grisham thriller *The Firm*. Although held up as gold standards for product placement in movies, these efforts pale in comparison to the sheer magnitude of product placement in movies of the past five or so years.

Take, for example, the hit movie *Marley & Me*, a real-life story about the life of newspaper man John Grogan and his lovable but mischievous Labrador, Marley. I won't give away the ending of this Owen Wilson and Jennifer Aniston flick, but there wasn't a dry eye in the theater. Despite the wonderful story, the movie's producers were mindful of the bottom line and featured the following brands during the course of the movie: AirTran, Apple, Banana Republic, Dell, Disney, Dos Equis, First Response, Graco, Honda, Huggies, IKEA, Kellogg's Corn Flakes, Lacoste, New Balance, *New York Times*, Perrier-Jouët, Philadelphia Eagles, Philips, Purina Dog Chow, *South Florida Sun-Sentinel*, The North Face, *Palm Beach Post*, *Philadelphia Inquirer*, Toyota, UPS, and Whirlpool.[25]

For those of you keeping track at home, that's twenty-seven product placements during *Marley & Me*. "Well," you say, "he just picked the most egregious example he could find." But you know better than that. According to an organization known as Brand Channel (brandchannel.com—could this be a product placement itself?), fifty-two films were number one for at least a week during the time period running from January 2007 through June 2008. In these fifty-two films, Brand Channel counted 1,251 brands, for an average of 24.1 brand placements per movie. This figure has remained very consistent since the new millennium.

Paul Blart: Mall Cop is an example of a movie on the higher end of the product placement spectrum. I watched this movie twice; it was both funny and touching. Blart wants to be a real cop, but his hypoglycemia keeps him from passing the required physical exam. Undaunted, he makes do as a Segway-driving (product placement?) security officer for the West Orange Pavilion Mall. The movie revolves as much around Officer Blart's personal life as it does his attempt to save the mall when it's taken hostage by a band of high-tech criminals. My favorite part of the movie is when Blart is having dinner with his mother (he lives at home, of course) and his daughter. As he's slathering peanut butter on his pie, he ruefully moans that it "helps fill the cracks of the heart" (loose translation).

Appealing though the movie is, it pulls at your purse strings as much as, or more than, your heartstrings. As best I could tell—and this is not a precise science—forty-nine brands appeared in *Paul Blart: Mall Cop*. That's a lot of brands competing for your attention in a movie whose running time is ninety-one minutes. So many brands competing with each other, and with the story itself—it makes you wonder how effective such placements are.[26]

Research into that very question by Moonhee Yang and David Roskos-Ewoldsen found that the level of "brand integration" impacted a product placement's effectiveness.[27] Yang and Roskos-Ewoldsen identified three levels of brand integration: (1) background, (2) used by main character, and (3) an integral part of the story. A good example of the third would be the use of the FedEx brand in the Tom Hanks movie *Cast Away*, where Hanks's character is a FedEx employee who crashes in a FedEx plane, and washed-to-shore FedEx packages help him survive. As you might have guessed, the more integrated the brand is into the story, the greater the positive impact it has on brand awareness and memory. Yang and Roskos-Ewoldsen's research also showed that brands are viewed more positively when they're used by the main character of a story. Finally, here's the kicker. In comparing moviegoers who had seen a particular movie (with a particular brand portrayal) against moviegoers who hadn't seen that movie (and thus hadn't seen the brand portrayal), they found that the first group of moviegoers were more likely to choose that brand than were moviegoers who hadn't seen the brand portrayal. When in doubt, I always rely on the following supposition: advertisers and marketers wouldn't pay good money for something that didn't work.

**PRODUCT PLACEMENT
IN MOVIES QUIZ**

See if you can identify the products
featured in the movies below.

1. In the 1986 hit *Top Gun,* fighter pilot Tom Cruise wore which brand
    of sunglasses?

    a. Oakley

    b. Armani

    c. Ray-Ban

    d. Kenneth Cole

2. The irascible alien E.T. in the movie of the same name, loved this
    brand of candy.

    a. Skittles

    b. Reese's Pieces

    c. Boston Baked Beans

    d. M&Ms

3. What is the traditional car for James Bond?

    a. Mustang

    b. Lamborghini

    c. Aston Martin

    d. Ferrari

4. What car brand is featured in the chase scene in *Matrix Reloaded?*

    a. Cadillac

    b. Mustang

    c. BMW

    d. Mercedes

5. What is the preferred candy for the title character in the *Hellboy* movies?

   a. Snickers

   b. Milky Way

   c. Baby Ruth

   d. Hershey's

6. Which cell phone brand was featured in *Star Trek: The Future Begins*?

   a. Samsung

   b. Motorola

   c. Nokia

   d. LG

7. What package delivery company was featured in *Cast Away*?

   a. UPS

   b. FedEx

   c. USPS

8. What car brand is almost exclusively used in *Transformers*?

   a. Ford

   b. Dodge

   c. GM

   d. Toyota

9. Which fast food chain was featured in *Demolition Man*?

   a. McDonald's

   b. Burger King

   c. KFC

   d. Taco Bell

10. Which car brand is featured in *Twister?*

    a. Ford

    b. GM

    c. Dodge

    d. Toyota

IF YOU ANSWERED . . .

1–3 questions correctly—you appear to eschew the materialistic values of the consumer culture or simply need to get a social life.

4–6 questions correctly—you are on the precipice looking down into the consumer chasm and can't decide whether to stay or to jump.

7–10 questions correctly—you are a card-carrying member of the consumer culture with all the rights and benefits associated with that distinction.

Quiz answers: 1c, 2b, 3c, 4a, 5c, 6c, 7b, 8c, 9d, 10c

As I mentioned before, product placement has made a comeback in television as well. Nielsen Media Research reports that there were 117,976 product placements across America's eleven most-watched TV channels in the first quarter of 2008 alone.[28] In 2010, the network TV shows with the most product placements included perennial favorites American Idol, The Biggest Loser, The Celebrity Apprentice, The Amazing Race 16 (that's right—16), and Dancing with the Stars.[29] During the broadcast of the 2011 Men's NCAA Championship basketball game thirty three brands either appeared or were mentioned on air—not counting commericials.[30] But it's not entirely the TV industry's fault. Thanks to new technologies in digital video recording (for example, TiVo) that allow consumers to zip through or zap commercials entirely, and the advent of video games and the Internet as alternative entertainment choices, advertisers have had to find new ways to get their message across to the consuming public.

Product placement offers several advantages to advertisers. As it relates to TV, viewers can't zap or bypass a product placement as they can a commercial; that means it's a good alternative to the typical thirty-second spot. Furthermore, product placement, whether in a movie or on television, carries an implied endorsement by the celebrity or actor using the product. Research has shown that this implied endorsement has a positive impact on viewers' attitude toward the product in question. Finally, audiences feel product placements add an element of realism to the movie or TV show. As long as such placements are not overdone (we may have already reached this point in both TV and movies) and are well integrated into the scene and story line, they have the potential to impact consumer awareness, attitudes, and—most importantly to advertisers—purchasing behavior.[31]

Today's prodigious use of product placements on TV was "inspired" by an episode of the popular reality show *Survivor* in its debut year. In this particular episode, the prize for one of the challenges the teams of contestants faced was a bag of Doritos and a Mountain Dew. The positive impact of sales on Doritos and Mountain Dew was so successful that a new era of product placement on TV was launched. Product placements are now staples on similar reality shows, including *The Apprentice*, *Keeping Up with the Kardashians*, *Kate Plus 8* (sorry Jon!), *America's Toughest Jobs*, *Cash Cab*, *The Real Housewives of Orange County*, *America's Next Top Model*, *Ice Road Truckers*, *Project Runway*—and the list goes on and on.

Product placement is by no means limited to movies and television shows, however. In fact, online computer games that promote specific brands are the fastest-growing segment of online entertainment—make that "advertainment," because their primary purpose is to promote products—and have actually surpassed total movie box office revenues! This is big business, and companies are well aware of it. The purpose of these "advergames" is well summarized by a company called YaYa, a pioneer in advergaming, which says it aims to "expose consumers to products by engaging them in an addictive game-play experience while simultaneously reinforcing a positive brand impression."[32]

To list all such games here would be pure folly, but one of my recent favorites was (these games have a short shelf life) Frito-Lay's Battle of the

Cheetos advergame. Combatants are invited to help wrest control of the web from the evil Surdawg of the Crunchy Alliance. Players can choose their army—Crunchy or Puffs—and help make the web safe again for Cheeto lovers everywhere. The advergame needs to be entertaining enough that players want to continue to play—and possibly e-mail it to friends and talk about it at school—but as its name suggests, its real purpose is to get consumers to interact with the brand. Currently, other popular advergames include Drawit (3M Post-it notes), Happiness Factory (Coke), Behind the Pizza (Dominos), Sneak King (Burger King), Coke Zero Playbook of Possibilities, and countless others; additionally, pretty much any new movie uses online games to spread the word and offer ample product placement opportunities.

The success of advergames speaks for itself. Retention rates are often ten times greater than for standard broadcast commercials. The distribution is impressive too: up to 50 percent of people who receive games via promotional e-mails play, and the average time spent is twenty-five minutes; they also talk to a lot of other people online and e-mail the game along, with 90 percent of those who receive these pass-along e-mails playing as well. Whether the advergame revolves around cars, movies, beer, or shampoo, it builds a relationship between consumers and a brand by transferring the positive emotion of the game to the brand it revolves around—something we call "associative advertising."[33]

Let's not forget plain old video games as a promotional channel. No longer relegated to adolescent and young adult males, video games are played by more than 122 million people (about 57 percent of the population). It's hard to imagine a home without an Xbox, Game Boy, or Wii console, making these and other gaming systems a viable advertising channel. About 44 percent of video game players are now female, which opens up many new opportunities for advertisers. Games such as iVillage, PopCap Games, and "The Sims" series are so popular that many ad agencies have video game divisions.[34]

Regrettably, and on a much smaller scale, even the sanctity of the publishing industry has been invaded by product placements. Tina Wells, the CEO of Buzz Marketing (whose mission is to help companies spread the word about their brands), obviously practices what she preaches. "What does this have to do with books?" you ask. Good question. Wells is also the author of the "Mackenzie Blue" series of books for young people, "which offers companies the opportunity to sponsor the books and have their product featured in the story." Bad press has Wells backtracking, however, exclaiming that future books will not contain paid product placements.

Consumer watchdogs have suspected for some time that a series of popular books developed by Alloy Entertainment, including best sellers *Gossip Girl*, *A-List*, and the "Clique" series, contain paid product placements. According to one study, each book has approximately one brand mention per page. The *Boston Globe* pointed out that at one time Alloy's website had this to say: "Advertisers have the opportunity to get their products or services cast in these best-selling books."[35]

In a recently published volume for teens called *Cathy's Book*, Procter & Gamble struck a deal with the book's authors to integrate several of its products for teenage girls into their book in exchange for Proctor and Gamble's marketing support. When word of that arrangement got out, the book was boycotted by a consumer watchdog group, and critical blogs and editorials railed against such practices. When the paperback version of the book was published, all such references to specific Cover Girl products were noticeably absent.[36] My eldest daughter, a fan of the "Clique" series, feels that the use of real brands adds authenticity to the books. Maybe it's her reading of such books that has led to her acute interest in fashion. Lately she's been referring to herself as a "fashion diva."

BREAKING THE "FOURTH WALL"

How would you feel if you found out that the cell phone recommendation you heard yesterday at the office was from a co-worker who, unbeknownst to you, was paid to spread the word about that particular brand of phone? Be prepared: it's happening all the time. Have you ever heard of the "fourth wall"? This is "a theater term used to describe the imaginary boundary between the stage and the audience."[37] Marketers have finally broken the fourth wall of commerce—the boundary that, until now, separated the world of marketing from people's private, day-to-day lives. Marketers have enlisted your friends, family members, and acquaintances as a stealth marketing force to sell you the latest goodies our consumer culture has to offer.

Word of mouth (WOM), often referred to as "buzz," entails the informal exchange of information among consumers about products and services. Today's media clutter and people's ambivalence, if not hostility, toward more traditional forms of advertising have given rise to the increasing use of word of mouth marketing (WOMM—note the extra M there). WOM—single M—is considered to be more credible than other information sources, such as advertising and salespeople, it's relatively low cost, and it's easily spread online. A recent study found that approximately two-thirds of all sales of consumer goods are based on WOM.[38]

It would be difficult to argue that WOM does not play an integral role in the consumer purchasing process. Purchasing is a social act that involves not only an interaction between a company and its consumers but also a complex exchange of information between each of us and those people who surround us. WOMM—double M now—simply builds on people's natural inclination to share their thoughts, feelings, and experiences with others. A recent study by the Keller Fay Group found that over 90 percent of all WOM conversations in the United States take place offline (with about 75 percent of the some 3.5 billion daily WOM conversations being face-to-face interactions and 17 percent occurring via the telephone), and only 7 percent occur online via e-mails, text and instant messages, blogs, and chat rooms. Approximately half of that online traffic consists of recommendations made by family and friends. Mass media channels like radio and television are better at teaching and informing about products, but it's interpersonal communications that are the most persuasive.[39]

Let's take a look at how WOMM works. At first glance there appears to be no rhyme or reason to the tsunami of never-ending consumer chatter that washes over us. Upon closer inspection, however, some patterns emerge. Consumers exhibit individual differences in their willingness to try and accept new products. Some of us want to be the first on the block to own a new product—we want to be at the vanguard of consumption—

while others wait for more information or rely on the experiences of people who are more adventurous before buying a product. Still others wait even longer, deliberately opting for what eventually is not such a new product. Some of us never buy certain products, or even categories of products.

Pioneering research by Everett M. Rogers, published in his now classic 1955 book, *Diffusion of Innovation*, helped explain how new ideas and products "diffuse" throughout the marketplace. From a marketer's perspective, "diffusion" is the process by which an innovation spreads to new consumers by various means of communication, including mass media, sales people, and WOM over a given period of time. Rogers identified five "adopter categories" based upon the length of time it takes consumers to adopt an innovation. His model, depicted below, is bell-shaped. What this means is that at first the adoption of any new product is slow, but adoption picks up speed as the product becomes more diffused throughout society; adoption then eventually tapers off, as fewer and fewer potential adopters remain.

Rogers's Adopter Categories

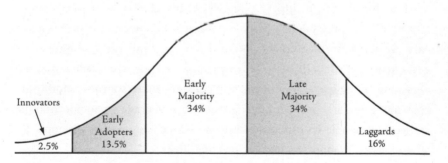

The first category of adopters, as you can see, is "innovators," who comprise only 2.5 percent of the population. They are adventurous, are willing to take risks, and tend to be younger, better-educated, in a higher income bracket, and self-reliant. Innovators don't look to others but prefer nonpersonal sources of information such as breaking research and experts in their fields of interest. They are highly cosmopolitan, maybe reading the *London Financial Times* or the *New York Times*, which gives them a rather global perspective.

Such individuals seek social relations outside their local peer circle and rely on their own judgment rather than group norms. Their adventurous nature and self-reliance often lead them to try new products that may be expensive and cutting edge. Someone in this group might have been the man or woman to buy a mobile phone when those early models were the size of a shoebox.

People in Rogers's second adopter category, "early adopters," comprise approximately 13.5 percent of the population and are much prized by marketers. Early adopters have extensive social networks and are deeply involved in their local communities and are often leaders in local clubs and organizations. These so-called opinion leaders are the primary target of WOMM. They are more sensitive to group norms and values than the previous category and are considered trendsetters. Unlike innovators, early adopters rely more on mass media and salespeople for product information and are heavy users in the product category of their expertise.

The third, "early majority" category makes up 34 percent of the population. This is pretty much Middle America. People in this group make careful, deliberate decisions. Unlike innovators and early adopters, the early majority is less likely to "throw caution to the wind" when making decisions. They spend more time and consider more sources before purchasing. The early majority is slightly above average in socioeconomic status, and people in this group rely on advertising and salespeople (as well as on early adopters) for new product ideas.

The "late majority" category also makes up 34 percent of the population. These folks are below average in income and education and are skeptical. They often lack the finances to purchase new products readily and must be convinced that any new product is worth the expense. Members of the late majority will spend their money but only after it's been shown to be a good value or after many of their friends and family have finally taken the plunge. They rely heavily on WOM communication and look to earlier adopters for clues on which products to buy.

"Laggards" comprise the final 16 percent of the population and occupy the lowest rungs on the socioeconomic ladder. They are somewhat disenfranchised and don't like change. Laggards are tradition-bound ("It's the way we've always done things") and are often driven by a lack of financial resources. When a laggard ultimately purchases a product, it is likely out of date already, with earlier adopter categories having already moved on to the next generation of the product. Laggards may be targets for companies trying to clear their shelves of products the rest of the market has already passed by.[40]

Marketers have long recognized the importance of innovative or adventurous consumers in gaining acceptance for new products. These consumers pass along information and make recommendations about new products to others—the linchpin of successful WOMM. Extensive research has been conducted to identify the profile and media habits of "opinion leaders" so that initial promotional efforts can be targeted to those particular consumers. The kicker is that someone who is an opinion leader in one product category—let's say fashion—may be a follower in another category—maybe computers.

Opinion leaders are knowledgeable and influential—people with large social networks who share advice with others. It is their position at the hub of a large social network and their willingness and desire to share their opinions with others that make them so valuable to WOMM. Opinion leaders are motivated primarily by a deep interest in a particular product category and an emotional need for attention. They like to be seen as a font of information and the center of their social solar system. The marketing premise is simple: convince these opinion leaders that your product is worthwhile, and they will spread this information to people throughout their vast social networks (who will in turn influence others in their social networks).

In and of itself, WOMM is not inherently evil. Like personal selling and advertising, it is a perfectly acceptable promotional tool when practiced ethically. Let's take a quick look at several firms' WOMM efforts and let you be the judge. The aptly named BzzAgent is a Boston-based

WOMM firm. Since its founding in 2001, BzzAgent has put together a nationwide WOM network of 60,000 "agents" in the United States. They are currently setting up networks across Canada and Great Britain as well. Founder and CEO Dave Balter's vision, according to the company's website (www.bzzagent.com), was to "create a company that would allow marketers to accelerate and measure honest word of mouth, while giving consumers a fun way to try new products and influence brands." Since its inception, BzzAgent has executed over 600 WOMM campaigns for hundreds of major brands, including Unilever, Kraft, Dunkin' Donuts, Philips, 7-Eleven, Cover Girl, and Neutrogena.

Unlike some of its competitors in the WOMM space, BzzAgent is not particularly picky as to who becomes a BzzAgent. In fact, anybody can sign up to be an agent on the company's website in just a few minutes. These agents run the gamut with respect to age, gender, income, and nationality. When BzzAgent gets a new client, the marketers put together a guide with the particulars of the product they will be promoting, including its major selling points and helpful hints on how to create "buzz" for the product. In addition to this, agents receive samples of the product being promoted and often coupons to distribute. They are then encouraged to sing the praises of the brand far and wide to friends, family members, co-workers, neighbors, current and potential retailers of the product, and just about anyone else who will listen, and to report back to BzzAgent about their efforts.[41] Much of the WOM marketing that agents undertake happens via face-to-face conversation. However, agents are also encouraged to harness the Internet and social networks such as Facebook, Twitter, and MySpace to spread the word. The use of any forum where a product mention or review might be seen by others is encouraged. BzzAgent has also launched a new service called BzzScapes, similar to other social networking sites. The aim of *this* one, however, is not to connect users with each other, but to allow users to interact with brands and to create "brand communities." Manufacturers follow these comments very closely, hoping to glean important insights from desired customer groups. For $5,000 a month ($3,500 with a six-month agreement), BzzAgent provides clients with trend reports and related personal profiles of agents.[42]

The strategy seems to be working. BzzAgents have shared the product gospel with over 100 million people and have submitted 2 million reports detailing "first-hand brand-related conversations" (company website). Even Balter is surprised that so many agents are willing to spend five to ten hours each week chatting up the benefits of, say, Dunkin' Donuts's new Latte Lite product for little remuneration other than product samples and coupons (though they can also earn points and collect rewards for their efforts). As the "recruitment" portion of the BzzAgent website suggests, agents like to be a part of a community of people changing the world, enjoy being the first to learn about new products and services, and appreciate always having something to talk about with their friends. It's all about building their social capital.[43]

For some people this raises a red flag. Journalist Rob Walker, writing for the *New York Times*, asks, "Do we really want a world where every conversation about a product might be secretly tied to a word-of-mouth campaign?" Doesn't that kind of undermine, you know, "the fabric of social discourse"?[44] And it is precisely here that such WOMM campaigns cross over to the dark side. Agents are not required to disclose their affiliation with BzzAgent or the brands they're promoting, and few do. I don't blame them. If they were to do so, they would be seen very differently—not as a friend sharing an honest opinion about a product, but as a paid (albeit lowly paid) shill for a company that accepts six-figure fees from corporations to market their products. Clearly, without the disclosure of their selling intent, such agents and the WOMM campaigns they're supporting break the above-mentioned "fourth wall" that protects our personal lives.

But it's not just the fourth wall that's being broken; it's also the industry's ethical standards. The 400-member Word of Mouth Marketing Association (WOMMA), the leading trade association for the WOMM industry, clearly discourages such deception in its code of ethics. Its guidelines regarding honesty and its fourth fundamental principle can be found below.[45]

WOMMA Code of Ethics

WOMMA FUNDAMENTAL PRINCIPLE 4: WORD OF MOUTH CANNOT BE FAKED

Deception, infiltration, dishonesty, shilling, and other attempts to manipulate consumers or the conversation are bad. Honest marketers do not do this, will not do this, and will get caught if they try. Sleazy behavior will be exposed by the public and backfire horribly on anyone who attempts it.

WOMMA ETHICS CODE 2: THE HONESTY ROI—HONESTY OF RELATIONSHIP, OPINION, AND IDENTITY

Honesty of Relationship

- We practice openness about the relationship between consumers and marketers. Consumers engaged in a word of mouth program should disclose their relationship with marketers in their communications with other consumers. We don't tell consumers specifically what to say, but we do instruct them to be open and honest about any relationship with a marketer and about any products or incentives that they may have received.

- We require marketers to disclose their relationships with consumers in relation to word of mouth initiatives.

- We require marketers to effectively monitor disclosure of consumers involved in their word of mouth initiatives.

- We stand against marketing practices whereby the consumer is paid cash by the manufacturer, supplier or one of their representatives to make recommendations, reviews or endorsements.

- We require consumers involved in a word of mouth initiative to disclose the material aspects of their commercial relationship with a marketer, including the specific type of any remuneration received.

- We require consumers involved in a word of mouth initiative to disclose the source of product samples or incentives received from a marketer.

- We comply with FTC regulations that state: "When there exists a connection between the endorser and the seller of the advertised product which might materially affect the weight or credibility of the endorsement (i.e., the connection is not reasonably expected by the audience), such connection must be fully disclosed."

Honesty of Opinion

- We never tell consumers what to say. People form their own honest opinions, and they decide what to tell others. We provide information, we empower them to share, and we facilitate the process—but the fundamental communication must be based on the consumers' personal beliefs.

- We comply with FTC regulations regarding testimonials and endorsements, specifically: "Endorsements must always reflect the honest opinions, findings, beliefs, or experience of the endorser. Furthermore, they may not contain any representations which would be deceptive, or could not be substantiated if made directly by the advertiser."

Honesty of Identity

- Clear disclosure of identity is vital to establishing trust and credibility. We do not blur identification in a manner that might confuse or mislead consumers as to the true identity of the

individual with whom they are communicating, or instruct or
imply that others should do so.

- Campaign organizers should monitor and enforce disclosure
  of identity. Manner of disclosure can be flexible, based on the
  context of the communication. Explicit disclosure is not
  required for an obviously fictional character, but is required for
  corporate representation that could be mistaken for an average
  consumer.

Only applicable portions of the WOMMA Code of Ethics are included above.

WOMMA's fourth fundamental principle, quoted here, is clear on the
subject of disclosure—deception and shilling (the practice of promoting a
product without others' knowledge) are explicitly discouraged. The principle is worth repeating: "Honest marketers do not do this, will not do
this, and will get caught if they try." Interestingly, BzzAgent founder Dave
Balter was a cofounder and is a current executive council member of
WOMMA. Clearly, someone is not practicing what he preaches.

Even a cursory glance at the above guidelines reveals that WOMMA
stands firmly behind full disclosure for those implementing WOMM
campaigns. Every paragraph but one under the "Honesty of Relationship" section calls for disclosure with other consumers. Brand marketers
are also encouraged to monitor whether their agents are disclosing their
relationship to WOMM campaigns when interacting with unwitting
consumers. The last bullet-point hits the nail on the head: when the relationship between an endorser and a seller might materially affect the endorsement's credibility, that connection must be fully disclosed.

The "Honesty of Identity" section further drives home the importance of disclosure of any relationship between an endorser and brand
marketers. Clear disclosure, it states, is "vital to establishing trust and
credibility." In a 2009 article by Kristen L. Smith, executive director of
WOMMA, she encourages/exhorts companies to be transparent and to
disclose. She states very clearly, "Don't engage in stealth marketing and

do not shill—i.e. pay people to talk about or promote your product without disclosing their relationship to you."[46]

Yet this is precisely what many companies are doing. And it's not just companies you've never heard of. Corporate behemoth Procter & Gamble (P&G), the maker of such consumer stalwarts as Dawn, Bounty, Pampers, Charmin, Old Spice, Pringles, and Crest toothpaste, to name but a few, has also embraced practices on the ethical fringe of WOMM. P&G is the only major corporation to have developed its own in-house business unit devoted to WOMM. In 2001 the company launched Tremor, a WOMM initiative focused on teens ages thirteen to nineteen. Through a website also called Tremor, a sales force of approximately 250,000 "connected" teens was recruited. Unlike BzzAgent, however, P&G carefully screens candidates and accepts only 10 to 15 percent of those who apply. Tremor CEO Steve Knox notes that while most teenagers have twenty-five or thirty names on their instant-messaging "buddy list," Tremor teens might have five or six times that. P&G is looking for only highly connected, trendsetting teens. Tremor teens get discounts, downloads, free samples, and coupons, and an occasional check for taking surveys and participating in focus groups.

Procter & Gamble does not tell "Tremor Crew" members what to say, but neither does it require them to disclose their membership to anyone—including their parents. Commercial Alert, a consumer activist organization, lodged a complaint with the Federal Trade Commission arguing that Tremor takes advantage of their crew members. Without disclosure, warns Commercial Alert cofounder and former executive director Gary Ruskin, we run the risk of the "commercialization of human relationships" and, as Robert Berner adds, we end up treating others as mere "advertising pawns, undercutting social trust." Others have argued that lack of disclosure of the teens' relationship with P&G and Tremor violates section 5 of the FTC act that prohibits "unfair or deceptive acts or practices." Paradoxically, P&G is a member of WOMMA, which—as we saw above—strongly encourages full disclosure of any relationship between a brand and its endorser.[47]

Given that most of P&G's products target adults, they launched another WOMM initiative called Vocalpoint. P&G has enlisted nearly

600,000 mothers, aged twenty-eight to forty-five from across the United States to participate in WOMM campaigns. Like their Tremor counterparts, Vocalpoint mothers are connected. While the average mom will talk to five women in a typical day, Vocalpoint moms speak with twenty-five or thirty, making them just the type of influentials (opinion leaders) needed to jump-start a new brand introduction. Vocalpoint moms are encouraged to spread the word about P&G brands such as Olay skin products, Cover Girl makeup, and Febreze air fresheners. Vocalpoint handles WOMM campaigns for non-P&G products as well. Vocalpoint raises the same ethical dilemma that Tremor does. Should people who are part of a WOMM campaign disclose their affiliation? Again, P&G leaves it up to their "connectors" to make that decision.[48]

It appears that the fourth wall has not only been breached but blown to pieces. WOM marketing, as practiced by many, has co-opted human relationships for the sake of the almighty dollar. WOM marketing, in and of itself, is not bad, but many of its current practitioners violate the principles that protect its legitimate use. The ongoing use of deception and lack of disclosure by WOM marketers will continue to erode the distinction between human interaction and marketing efforts and further the commoditization of human relationships.

But all is not lost! The next three chapters on self-control offer glimmers of hope in our efforts to fight the marketing machine.

THE THREE INGREDIENTS
OF SELF-CONTROL

What lies ahead of us and what lies behind us is nothing
compared to what lies within us.

—RALPH WALDO EMERSON

Has it come to this? Children playing the role of parent, complaining
about Mom or Dad's excessive e-mailing? With a dizzying array of hand-
held electronic gadgetry available to check e-mails, notes *Wall Street
Journal* reporter Katherine Rosman, BlackBerrys and other e-mail-capable
phones are creating a rift in families.

The word "CrackBerry"—a nickname for the BlackBerry, reflecting
the device's addictive quality—has become a regular part of many people's vocabulary. Parents have taken to sneaking around the house to
check for and send e-mail messages, routinely lying to their children
rather than fessing up. Some have resorted to hiding their BlackBerry in
the bathroom or other "safe" spots around the house.

Many families, at the impetus of the youngsters, have had to establish
rules to minimize the intrusion such devices have on what was once "family time"—rules similar to those that parents used to impose on children's
computer use or gaming. Dinnertime, many parents now feel, is a good

opportunity to check for messages. Children tend to disagree. They don't like message-checking at other times either: many worry that their parents might have an accident caused by texting while driving; others complain that their parents check and send e-mails at school functions and at movies. Apparently no place is safe from such obsessive behavior—not even weddings and funerals! Many mental health professionals report that the intrusion of e-mail devices into the inner sanctum of family life is a much-discussed topic in therapy sessions. Children don't like taking a backseat to a parent's BlackBerry. They want (and feel that they deserve) undivided attention, yet they can see where they rank in parental priorities when a parent flaunts established rules about e-mailing.

Finding the proper balance can be difficult, and an imbalance is serious. People who have to check for messages every five minutes, and who feel anxious after a longer time span, may actually have an obsessive-compulsive disorder. Checking messages brings such people a small measure of relief, but shortly thereafter they experience a rising uneasiness that they might have missed something important. This problem has become so commonplace that children who have to compete with parental handheld devices are often called "BlackBerry orphans."[1]

WE ALL STRUGGLE WITH ISSUES RELATED TO SELF-CONTROL IN ONE AREA OF life or another. In the financial realm, the stakes are high. A lack of fiscal self-control keeps us off the road to happiness. Although Madison Avenue tells us otherwise, happiness cannot be purchased at the mall, on the Internet, or from a catalog. We're happiest when we feel good about who we are, have meaningful relationships with others, and get involved in our community—and those things can't come about unless we exercise financial self-control. It was my research on the role that self-control plays in how we spend our money that prompted me to write this book.

Did you know that the personal savings rate in the United States has been steadily declining for years? We have morphed from a nation of savers into a nation of spenders. A few years ago the savings rate for those past the Baby Boom—people forty-two years of age or younger in 2006—hit an all-time low of minus 18 percent. That's not a typo! Young people are spending quite a bit more than they are making. This is chilling given that most analysts recommend we save 15 to 20 percent of our pretax income for retirement or other goals. It has become so bad that many companies are automatically enrolling new employees in 401(k)s and withholding a bit of their pay for investing without even asking. Because young people can't resist spending what they earn, companies are taking the decision out of their hands.[2]

This picture is nearly as bleak for older Americans. Personal savings rates are near or below zero, resulting in little money for emergencies, children's college, or looming retirement. The typical Baby Boomer has about $50,000 in savings as he or she heads into retirement—woefully inadequate.[3] "Wait a second," you might say. "Hasn't the savings rate jumped as high as 5 or 6 percent since the recession of 2008–2009?" And you would be correct. But this doesn't mean that Americans have made a 180-degree turn on savings. A nearly 10 percent unemployment rate, the all-too-likely prospect of losing our job, and the barrage of bleak economic forecasts have us running scared. The rainy day is here *now*, and we were caught without an umbrella.

Unfortunately, in this instance I think that the past is the best predictor of the future. What our past tells us, as we saw in an earlier chapter, is that Americans quickly return to their profligate ways once a recession fades. Recessions in the 1970s, in the 1980s, and at the turn of this century (when the Internet bubble burst) have all been followed by a marked uptick in spending and a further erosion of any incipient savings ethic. Without a real change in our collective attitude toward savings and spending, history will likely repeat itself. But it doesn't have to for you. You can learn to apply the lost art of self-control to your finances and bring about a higher quality of life (not just financially but overall) for you and your loved ones.

THE LOST ART OF SELF-CONTROL

When we exercise self-control, we act in a way that is far-sighted: we think of the long-term effects of our actions rather than the immediate, short-term pleasures. Doing so often means controlling our emotions and our urges so that we can then control our actions. Self-control enables us to override our impulses and stop short of acting on them.[4]

There's a reason we refer to "exercising" self-control: it involves exertion. How much exertion it takes varies from person to person, however. Some darn people just have more control than others, though a person's ability to control him- or herself is pretty consistent across a variety of situations and stages of life. That consistency was revealed in the results of a two-part experiment in which (1) four-year-olds were given the choice of eating one marshmallow immediately or waiting a few minutes and getting two and then (2) those same participants were assessed as to their scores on college entrance exams over a decade later. What's amazing is that those children who had the self-control to wait for the two marshmallow payoff did better on their college entrance exams years later than their four-year-old counterparts who took the single marshmallow and ran.[5] This consistency suggests that genetics plays a role in our ability to control ourselves. Crucial to the premise of this book, however, is the belief that our ability to exercise self-control can be improved through practice. I'll share specific suggestions in chapters 13 and 14.

Just as regular exercise can increase our stamina, self-control can be improved the more we use it. This may explain why some people eventually find the strength to avoid overeating or shopping sprees after numerous prior attempts have failed. The good news is that we can break those old habits and establish new, healthy ones, but we must always guard ourselves against relapses. It truly is "one day at a time"; that mantra isn't just for recovering alcoholics or drug abusers, but for all of us as we fight to lead healthy and happy lives in a society that has given in to the shiny-objects ethos.

So how's your self-control? I have placed in the appendix at the end of this book a commonly used scale of self-control, along with norms for the United States. Take a moment to complete the scale and see if you're a paragon of self-control or, like the rest of us humans, have room for improvement.

Even if you *are* a paragon, you're not perfect. Self-control is a limited resource, and thus you can become exhausted from too many demands at one time. You may fail at self-control on a grueling day when you normally wouldn't under less demanding circumstances, because exertion leads to fatigue. Performing any act of self-control (even simply making a decision) appears to deplete a person's store of control. It's on those particularly bad days that you are most likely to cheat on your diet, skip exercising, argue with others, or go on a spending spree.[6]

Think of self-control as operating in a feedback loop, commonly called a "TOTE loop." TOTE is an acronym for test-operate-test-exit. A good example of such a loop is your typical interaction with the thermostat that controls the temperature of your home. As the temperature deviates from your desired setting, the situation is detected and the ambient air is either cooled or warmed. The first T in the TOTE loop—"test"—involves monitoring your current condition as opposed to your desired state. If you feel that you're not as comfortable as you would like, you move into the O stage—"operate"—and change the temperature setting. The operation stage is followed by the second T in the TOTE loop as you "test" to see whether the particular operation undertaken achieved your desired goal. If not, you continue to change ("operate") the temperature until it's just right—nice and cozy. If resetting the thermostat brings about the desired comfort level, you "exit" the loop.

If we assume that self-control operates in a feedback (TOTE) loop like the one above, there are three major ingredients needed to effectively cook up sufficient self-control: self-monitoring, clear standards, and the capacity to make changes in your thoughts and behavior.[7] Although you might expect the setting of standards to be the first ingredient, I'm going to start with self-monitoring, because only if you're monitoring your spending behavior will you be able to identify areas of concern; from there you can then set standards toward which to work.

INGREDIENT 1: MONITORING YOURSELF

Understanding and identifying the causes of self-control failure—in other words, doing self-monitoring—is the first ingredient needed to improve your self-control. Self-monitoring is the process of observing your environment and behavior and systematically noting (and preferably re-cording) relevant events and behaviors. Identifying a "target" behavior (the behavior you want to either limit or encourage) and recording how often, how long, and when the behavior occurs helps you get a clearer grasp on the frequency, timing, causes, and effects of that behavior. Self-control strategies place primary importance on self-observation as a cru-cial step in self-change programs. You must know what's happening before you can implement any self-change program. The good news is that the simple act of observing and recording your behavior may itself lead to behavior change; in other words, being aware of yourself and your environment is sometimes enough to bring about desired changes.

When you *don't* monitor your behavior or your monitoring system fails, self-control becomes difficult if not impossible. A failure to monitor may be the main reason that so many of us can't seem to follow through even when we desperately want to quit drinking or smoking, eat less, exercise more, or save money. One of the main problems with self-monitoring is that people don't do it enough. We call this "under-regulation."[8] Spend-ing money is a good example. When people don't keep track of their ex-penditures, their spending is likely to exceed what they can afford. Remember the earlier research which found that when we use credit cards, we are more willing to spend and are willing to pay more for items? The main reason for that is that we don't keep as close an eye on our spending when we use credit cards—in other words, less monitoring. With cash, checks, and debit cards, when we spend money it is no longer available to us. However, with credit cards the impact on our bottom line is delayed. This delayed impact on our wealth, along with the fact that most of us don't write down the amounts we charge, leads us to overesti-mate the amount of money we have. Thinking we have more money than we actually do increases the likelihood of future purchases.

Self-awareness can be painful for some people. To avoid feelings of inadequacy or personal hurt or loss, those people may attempt to narrow their attention to a single element in their life. In essence, they put on blinders to block out everything that might cause them pain. These blind-ers, however, can lead to a focus on relatively meaningless pursuits, such as shopping. Shopping as avoidance can become all-consuming, as we saw earlier, with no regard given to the long-term negative consequences associated with overspending: credit card and other debt, marital problems, and feelings of shame, guilt, remorse, and hopelessness. Instead of thinking with a long-term outlook, we give in to our impulses, which exist in the here and now and offer short-term pleasure at the expense of longer-term, more meaningful pursuits. Attending to your self, even when you don't like what you see, is the essence of the monitoring function.[9]

Monitoring is critical to good self-control, but it's not easy. Most of us are aware of the bad decisions that can be made when drinking alcohol, yet we continue to enjoy our cocktails. Why do you think this is so? It's because alcohol interferes with our ability (and desire) to monitor our behavior. When we become less self-aware, we operate impulsively, which leads to poor decision making and regret. However, alcohol is not the only culprit that undermines our ability to self-monitor. When we're under stress, distracted, in a bad mood, or preoccupied, it is less likely that we will do adequate self-monitoring.

The ABCs of Your Behavior

I mentioned that self-monitoring requires *noting* and even *recording* your behavior (and relevant events surrounding your behavior) for purposes of both feedback and evaluation. Not many people are in the habit of jotting notes about their own behavior. In essence, you must become a personal scientist, collecting relevant information about yourself and your environment that will be essential in choosing the methods needed to change your behavior. But how is such data collected?

Let's go back to the ABCs: think of your **b**ehavior as being "sandwiched" between its **a**ntecedents (those things that occur before it and are "cues") and its **c**onsequences (things that occur after, and because of, the behavior). When you have knowledge of and control over the antecedents and consequences of your target behavior, you can be incredibly successful at making desired changes, whether you're trying to limit or to encourage a particular behavior.[10]

There are many varieties of self-monitoring devices that can be used to keep track of target behavior, all of which require little or no monetary investment. Possibly the simplest is some form of a personal behavioral diary in which you can record (1) how often the behavior occurs, (2) what you think prompted the behavior, and (3) any resulting consequences. For example, if you suspect that you spend too much money shopping online, you need to track the exact number and time of day of your various expenditures. In addition, ask yourself, each time, what prompted you both to go online and to make a purchase. Were you bored? Anxious? Or were you checking e-mails when an ad caught your eye? Do you spend more at certain times of the day or week? Clearly, answers to such questions are invaluable in your efforts to cut back your online spending.

Other self-monitoring devices include small counters that enable you to record the number of times a behavior or thought occurs. These counters are often sold for other purposes (for example, counting golf strokes), but they do a great job as data collection devices in the financial area as well. Be creative: anything that will help you accurately and easily measure the target behavior will help you become more self-aware. Portable timing devices (stopwatches, watches with timers, cell phones, and so on) can also be useful, depending on the behavior to be measured.

Another option is a behavior graph or chart. Whether it's tracking your weight, how many cigarettes you've smoked, the miles you've walked, or the money you've saved for your retirement or the children's college, a chart or graph is a great way to monitor your progress. Be sure to place it in a prominent location that you (and possibly others) are sure to see repeatedly throughout the day.

As I noted earlier, simply keeping track of your behavior (how much money you spend, how you feel before going on a shopping spree) may result in improvement in that behavior.[11] However, don't be fooled into thinking that recording your behavior, on its own, is a surefire way to better self-control. Even with adequate monitoring, lapses in self-control may occur if you have conflicting goals and insufficient strength to resist impulses—in other words, if you can't forgo short-term rewards for larger but more distant payoffs (that is, can't delay gratification).

Measuring Self-Monitoring: How Conscientious Are You?

We will talk about conflicting goals and changing your behavior in the next two sections of this chapter. Before we do, though, let's see—using a short conscientiousness scale—how closely you monitor your thoughts, behavior, and surroundings.[12]

Conscientiousness Scale

Read each of the ten statements below and circle the extent to which you agree or disagree with each. If you are honest in your responses, this scale may help you to better understand the source of your lapses in self-control.

	Strongly disagree	Disagree	Neutral	Agree	Strongly agree
1. I am always prepared.	1	2	3	4	5
2. I pay attention to details.	1	2	3	4	5
3. I get chores done right away.	1	2	3	4	5
4. I like order.	1	2	3	4	5
5. I follow a schedule.	1	2	3	4	5
6. I am exacting in my work.	1	2	3	4	5
7. I leave my belongings around.	1	2	3	4	5
8. I make a mess of things.	1	2	3	4	5

	Strongly disagree	Disagree	Neutral	Agree	Strongly agree
9. I often forget to put things back in their proper place.	1	2	3	4	5
10. I shirk my duties.	1	2	3	4	5

SCORING INSTRUCTIONS

1. *Add up your scores for items 1–6. Your subtotal = _____*

2. *Reverse your scores for items 7–10 and then add them. That is, if you scored a 5 give yourself a 1; if you scored a 4, give yourself a 2; and so on. Your subtotal = _____*

3. *Add your subtotals for steps 1 and 2 and divide by 10 (the number of questions in the scale). Your conscientiousness score should range between 1 and 5.*

Write your conscientiousness score here: _____

Based upon a recent sample of 2,663 college students and numerous other research studies investigating conscientiousness over the past twenty years, several benchmarks are available for seeing how you rate compared to others on this scale. An average conscientiousness score across the above-mentioned studies is about 3.5. That means if you scored above 3.5 you are in the top 50 percent of the population in regard to conscientiousness. If you scored above 4.1, you are very conscientious. Above 4.7 and you may need to relax a little. On the other hand, if you scored below 3.5 you are in the bottom 50 percent of the population in regard to conscientiousness. Below 2.9 and you are probably struggling to keep your life in order. If you scored below 2.3, I'm surprised you were able to find a pencil with which to complete the scale!

In all seriousness, your conscientiousness score should give you a good idea whether keeping (or not keeping) track of your behavior might be contributing to any self-control issues you're experiencing.

INGREDIENT 2: SETTING CLEAR STANDARDS

The second ingredient of self-control is standards, which can take many forms. Standards may be social norms ("More is better"), personal goals ("I want to be rich"), or the expectations of others ("I always thought he/

she would be more successful"). When standards are lacking, are ambiguous, or contradict each other (the latter being particularly relevant for our discussion), self-control failures are more likely. One of my favorite sayings, used as the epigraph to chapter 7, goes like this: "If you don't know where you're going, any road will get you there."

Clear standards are just as critical to your self-control as a planned destination is to your journey.[13] In tandem with self-monitoring, standards provide a benchmark for assessing improvement in your behavior. If you have standards that conflict with each other, so that progress toward one goal means distancing yourself from another, you've got problems. For example, if you're highly materialistic but also want to be a good parent and be involved in church and community activities, you're likely to experience a lot of stress and anxiety in trying to accomplish these conflicting goals. Your materialistic pursuits require long hours at work (and time spent acquiring and using the spoils of that work)—all of which comes at the expense of family time, personal growth, and community involvement. In this respect life is a zero-sum game: time spent on one activity takes away time available for other pursuits.

Recent research has found that conflicting values create mental stress as we struggle to find the proper balance, which in turn leads to lower well-being and increased psychological problems such as depression and neuroticism. This condition is a serious concern, since the tension created by conflicting goals is likely to be both a daily battle and a life-long struggle. Chronic stress makes those affected particularly sensitive to impulse buying in a consumer culture that tries to convince us that happiness can be bought. Additionally, stress can make people more susceptible to influence from advertising, pushy salespeople, and other marketing efforts.[14]

In addition to inhibiting self-control and leading to impulsive behavior, conflicting or absent standards make it difficult for people to manage their lives. They find it increasingly difficult to make decisions, until finally they are paralyzed by doubt. Too much time is spent thinking about options and priorities and too little on action. The consequence of such decision paralysis is that little or no progress is made toward any goals.

A state of conflict in standards is quickly becoming the rule rather than the exception. Nowhere is this truer than in the realm of material possessions. We're up against a multi-billion-dollar advertising industry whose sole purpose is to get us to spend our money, and they're good at it. We're constantly bombarded with the message that we don't quite measure up. Madison Avenue's primary strategy is to convince us that if we buy this product or that, we will become the people we want to be and will achieve the happiness that has always eluded us. You know as well as I do that this just isn't true. No product will make you happy for more than a fleeting moment—not even that $2,000 Aura cell phone—if you're not satisfied with who you are on the inside.

As detailed earlier, numerous research studies have found that highly materialistic people are less happy and are more stressed and depressed than less materialistic people. Moreover, materialistic people report being sick more often than their less materialistic counterparts. As humans, we have an innate need to feel good about ourselves, have close relationships, and have a sense of being a member of a community. The pursuit of material possessions has the opposite effect. Rather than bringing us well-being, it alienates us from our friends and relatives, encourages us to favor possessions over relationships, and reduces the likelihood that we will donate money to charity, get involved in our community, or participate in public issues such as environmental protection, school quality, and crime prevention. In essence, the pursuit of materialistic goals keeps us from the very things that bring true happiness and fulfillment.

Emotional distress, including the distress resulting from materialism and from conflicting goals, makes people focus on the short-term goal of improving how they feel at the moment at the expense of more rewarding long-term goals. For example, although you may place a premium on saving for retirement or a child's college education, the immediate desire to improve your mood takes priority and impulsive behavior may occur. This situation may be particularly serious when you use shopping as a form of entertainment or mood repair ("retail therapy"), or go to the mall with no particular purchase in mind.[15] That's why clear standards are so important: those retirement savings or other long-term goals have to be kept front and center at all times.

Conflicting Financial Values: How Do You Measure Up?

Take a moment to assess how conflicted you are in regard to financial matters. Shown below is a short scale that measures how consistent your values are in regard to spending and saving money.[16]

Conflicting Values Scale

Read each statement and respond by circling the number that best represents how well a particular statement reflects the values you hold. Again, be brutally truthful in your responses. As a self-revelatory exercise, this can be helpful only if you're honest.

Assume, for the purpose of this exercise, that you are shopping in a store and find something that you really want to buy but that you really can't afford. Using a scale from 1 to 5 (where 1 is "Strongly disagree" and 5 is "Strongly agree"), indicate how much you agree or disagree with each statement regarding your feelings about whether or not to buy this item.

	Strongly disagree	Disagree	Neutral	Agree	Strongly agree
1. This would be a clear decision for me.	1	2	3	4	5
2. I would be conflicted between what I felt I should do and what I would like to do.	1	2	3	4	5
3. This would be a difficult decision for me.	1	2	3	4	5

	Strongly disagree	Disagree	Neutral	Agree	Strongly agree
4. I would feel a sense of conflict between my values in trying to make a decision.	1	2	3	4	5
5. There would be *no* wavering on which direction I would go.	1	2	3	4	5
6. I would *not* feel any internal conflict about my decision.	1	2	3	4	5
7. I would probably look back and wonder if I made the right decision.	1	2	3	4	5
8. I would feel somewhat guilty about the thoughts I had while working through my decision.	1	2	3	4	5
9. This is a situation for which I would need more information before reaching a decision.	1	2	3	4	5

SCORING INSTRUCTIONS

1. *Add up your scores for items 2, 3, 4, 7, 8, and 9. Your subtotal: _____*

2. *Reverse your score for items 1, 5, and 6. That is, if you scored a 5, give yourself a 1; if you scored a 4, give yourself a 2; and so on. Your subtotal: _____*

3. *Add your subtotals from steps 1 and 2. This total represents your conflicting values score and should range between 9 and 45.*

Write your conflicting values score here: _____

The benchmarks below are based upon a recent sample of 403 adults from across the United States ranging in age from eighteen to sixty-five who completed this conflicting values scale as part of a larger survey. Let's see how you stack up against them. A high score is worse than a low score: it means you feel more conflict when it comes to spending and saving money.[17]

Our sample of adults averaged 25 on the conflicting values scale. If you scored below 25, you're in the bottom 50 percent of the sample in terms of conflicting monetary values. This is a good thing: you're at peace at least when it comes to spending money you don't have. If you scored below 20 you are in the bottom 25 percent of the sample. This means you are among the elite in terms of having healthy values regarding money. However, if you scored above 25, you struggle with money issues more than half of the sample does; and if you scored above 30 your values conflict more than 75 percent of other shoppers in this country.

I won't sugarcoat it: setting and maintaining goals that focus on self-acceptance, personal relationships, and community involvement isn't easy. The pursuit of "stuff" is so pervasive, the advertising barrage encouraging materialism so constant, that it takes a person with clear and consistent standards to overcome such forces. And, as noted earlier, you have to take it "one day at a time." But if you can get off the consumption treadmill—and you can!—you will find that step life-giving and liberating.

A few years ago I was interviewed about the voluntary simplicity movement for the *USA Today* newspaper.[18] As part of this story, reporter Elizabeth Weise also interviewed a small group of friends in San Francisco, California, who had made a pact to buy nothing new except food, medicine, and toiletries for six months. The group called itself "The Compact" after the Mayflower Compact of 1620, which was a civil agreement among the Pilgrims to live a life of "higher purpose."

Rob Picciotto, a member of the twenty-first-century Compact, explains that the original purpose of the new Compact was not to save money or even the environment, but to simplify the members' lives. And it worked. "It saved us time because there was less time spent shopping.

We still buy groceries and go to the drugstore, but we don't go to Target on a Saturday, which was a ritual before just to see what the sales were," notes Picciotto.[19] Many of us don't realize (or even wonder) how much of our time is consumed by shopping—time that could be put to better use elsewhere: spending a day at the zoo with the family, reading a book, getting exercise, volunteering in the community, or getting some much-needed rest and relaxation.

Here's another example. Forced to cut back because of the recession, the Muniz family of Lyndhurst, New Jersey, found happiness in simplicity. Before the hard times hit, Al Muniz's jobs in car sales and home reconstruction kept him away from his family a lot. "In the old days, Al was always working," notes his wife, Angela. Family nights out at a restaurant were regularly interrupted by calls to Al requesting that he help solve the latest crisis at work—but no more. Now dinners at fancy restaurants have been replaced with a picnic or a walk on the beach. "I'm not mad at him all the time," says Angela, and she adds, "We've been much happier since he slowed down."[20]

INGREDIENT 3: DEVELOPING THE CAPACITY TO CHANGE

Mustering the willpower to resist temptation is at the heart of attempts at self-control. Even with clear goals and close monitoring of your situation, you need strength to resist life's many guilty pleasures. As we saw earlier, self-regulation is a limited resource, and failure may occur due to an insufficient supply of this resource at the time it is needed. Not having enough strength to combat one's impulses is often called "ego depletion." Less scientific terms include adjectives such as "worn out," "stressed," "frazzled," "at wits' end, and "dog-tired."[21]

You can become ego-depleted for many reasons: hectic schedule, lack of sleep, coping with negative life events, work complications, and the minor stresses and strains of everyday life. Any event that causes ego depletion contributes to self-control failures, but being physically tired is one of the worst offenders. The resources needed for self-control are restored during sleep and depleted yet again as the day progresses—some days faster than others. That's why so many self-control transgressions occur at night or at the end of a long week. The same holds true for smoking, drinking, drug abuse, sexual infidelities, and even criminal acts. Although this explanation isn't an excuse for impulsive behavior, understanding how (and when) lapses can be expected helps us better avoid them. Maybe Mom and Dad really knew what they were talking about when they told us to "sleep on it" before making important decisions: we may go to bed with a seemingly insurmountable problem, only to wake up feeling refreshed with a new outlook on life.

I use this very advice—"sleep on it"—when counseling on how to avoid credit card abuse. If you see something you really want to buy, simply have the salesperson place the item on hold until the next day. This gives you the opportunity to avoid an impulse purchase and instead to make the purchase decision the next morning, when your self-control resources are at their strongest. If you're like me, 90 percent of the time you won't go back and make the purchase. Hey, this self-control stuff might work!

Stress is almost as bad as fatigue at causing self-control failures. When you become "stressed out," you're more emotional and unstable than usual, and thus are more prone to take that short-term approach to decision making that we talked about earlier. You're more likely not just to spend money rashly, but also to participate in the many other guilty pleasures of life—and then, of course, you experience the negative outcomes associated with those pleasures. Making yourself feel better at the expense of longer-term concerns—we call this "emotion-focused coping"—leads to various breaches in your self-control. Worse yet, when you wake up the next morning, whatever originally caused the lapse in your self-control is still there, but you now have the added baggage that comes with your self-control failure!

Relaxation Techniques

Given that stress plays an important role in our ability to exercise self-control, we need to make relaxing a part of our daily routine. For many of us, this requires a big lifestyle change. We need to slow down and relax by walking, talking, eating slowly, smiling, laughing at ourselves, admitting mistakes, taking time to enjoy life, and renewing our faith. I encourage you to try as many of the following replenishing techniques as possible:[22]

1. **Plenty of sleep.** As we just saw, people can better handle life's many crises when they have gotten a good night's sleep. There's a reason the old adage to "sleep on it" is still around. It's good advice for potential spenders, because the resources needed to exert control over behavior are replenished with sleep.

2. **Aerobic exercise.** Prolonged exercise of at least 20 minutes is great for both heart and lungs and is a great stress reducer as well. It also works to lower depression, in many instances. The key is to choose something you enjoy doing and make it part of your everyday routine. I like to take walks, but you can jog, play a round of golf or take a Zumba class. Start slowly and be sure to reward yourself for exercising.

3. **Meditation.** A quiet time when the mind is encouraged to block out the stress and anxiety of daily life—meditation, in other words—has been shown to be a great relaxation technique. Quietly say a soothing word as you inhale and another when you exhale. Set aside fifteen to thirty minutes each day, maybe in the middle or toward the end of the day (when self-control lapses become more of an issue), and simply *be*.

4. **Progressive muscle relaxation.** If you systematically tense and relax different muscles, eventually they (and you) are relaxed. Start at your feet and work your way up. By narrowing your focus to your body's response to stressors, you block out the thoughts or events causing the tension. It's a great technique that can be used along with or as part of your meditation ritual.

5. **Visualization.** Believe it or not, you can use your imagination to relax and reduce tension. Through the process known as "visualization," you can create positive mental images that bring a smile to your face— maybe a favorite vacation spot or some place you would like to visit, a person or event in your life that brings a smile to your face, a positive image of you succeeding at an important task, or anything else that conjures up pleasant thoughts. You can use a favorite image each time you do this or mix it up depending upon current circumstances.

6. **Massage.** Nothing relaxes like a good massage. Although my favorite is having a significant other do the massage, massage therapists are also wonderful—but expensive. Self-massage is a distant third choice, but it can do the trick in a pinch.

7. **Social support.** Our relationships with others are the essence of life. Having supportive friends and family may be the Holy Grail of contentment and happiness. Having close friends makes us healthier and helps us live longer; likewise, people who are married live longer, healthier lives than their unmarried counterparts.[23] Having a loving and supportive network of friends and family helps buffer us from the stresses and strains we inevitably confront.

8. **Spirituality and faith.** The power of prayer and faith in the existence of a higher being has allowed many of us to overcome life's most difficult challenges. Such difficulties may have seemed insurmountable if not for our faith—and indeed, they may have *been* so. As a predictor of health and longevity, religious involvement rivals nonsmoking and exercise. In fact, religious attendance can lengthen your life-span by up to eight years. That's because churchgoers (a) exhibit more healthy behaviors (less smoking and drinking, for example), (b) enjoy the benefit of a strong social support system, and (c) experience more positive emotions, such as hope and optimism.[24]

Choosing Not to Make a Choice

As I have stressed repeatedly, performing an act of self-control depletes a crucial resource within you, temporarily diminishing the amount of that resource available to help you overcome other temptations at that time or in the future.[25] Now take that point and add a little twist: recent research has found that the same resource used in self-control—the one that helps us resist temptation—is also depleted through the process of making choices and decisions (whether related to shopping or otherwise).[26] Thus all of the decisions we make early in the day deplete the resource needed for us to exercise self-control later in the day. It stands to reason, then, that the fewer times we face temptation and have to make a choice, the more potential for self-control we conserve. That means we can develop our capacity for self-control by choosing not to make so many choices—in other words, by limiting the number of times we put ourselves into temptation's way.

This is why humans are so prone to habits and other stable patterns of behavior. Having routines helps us avoid making so many decisions that our self-control resource is spent. By simplifying life, we save that resource for the more important willpower struggles. If everything we did involved making a well-thought-out, reasoned decision, we would be overwhelmed. When you drive home the same way each day, eat at your favorite restaurant again and again, or consistently buy the same brands at the supermarket or mall, whether you know it or not you're conserving your self-control resource.

Are You Up to the Fight? Measuring How Depleted You Are

Even though we all experience ups and downs in our store of self-control, chances are there's a level of control-readiness that's typical for *you*. Let's see just where that level is. The following short scale will measure how ego-depleted—how tired, frazzled, whipped, worn out—you are.[27]

Ego-Depletion Scale

Read each statement and respond by circling the number that best represents how you currently feel. Be honest in your responses.

	Never	Rarely	Sometimes	Often	Very Often
1. I feel tired.	1	2	3	4	5
2. I feel energetic.	1	2	3	4	5
3. I feel fit.	1	2	3	4	5

	Never	Rarely	Sometimes	Often	Very Often
4. I feel drowsy.	1	2	3	4	5
5. I don't feel clear-headed.	1	2	3	4	5
6. I feel exhausted.	1	2	3	4	5
7. I don't feel like doing anything.	1	2	3	4	5
8. I have the feeling I can handle the world.	1	2	3	4	5

SCORING INSTRUCTIONS

1. *Add up your scores for items 1 and 4 through 7. Your subtotal:* _____

2. *Reverse your score for items 2, 3, and 8. That is, if you scored a 5, give yourself a 1; if you scored a 4, give yourself a 2; and so on. Your subtotal:* _____

3. *Add your subtotals from steps 1 and 2. This total represents your ego depletion score and should range between 8 and 40.*

Write your ego depletion score here: _____

Again, based upon a recent sample of 403 adults from across the United States (ranging in age from eighteen to sixty-five) who completed this ego-depletion scale as part of a larger survey, here are the benchmarks for seeing how you stack up.[28]

Our sample of adults averaged 23.5 on the ego depletion scale. The higher your score, the more stressed out you are. If you scored 23.5 or below, you are in the bottom 50 percent of the sample in regard to ego depletion. If you scored 18 or below, you must be a Buddhist monk—very low reported stress. On the other end of the spectrum, if you scored over 26 you're in the top 25 percent of the sample, suggesting that you're feeling significant mental and physical distress. Your ego depletion score gives you a good idea whether relaxing and getting some rest may be the key to improving your self-control.

The present chapter argued the importance of self-control in controlling your behavior and explained the ingredients needed to exercise that control. Chapter 13 will now provide twenty-five techniques you can use to tweak your environment to help curb your materialistic longings.

STEP AWAY FROM THE SHOPPING CART: ENVIRONMENTAL PROGRAMMING FOR CONSUMERS

Those who flee temptation generally leave a forwarding address.

—LANE OLINGHOUSE

Even the Amish appear to have been swept up in the rising tide of materialism. As reported in the *Wall Street Journal*, a recent run on an Amish-owned bank in northern Indiana is proof of the hypnotic allure of the consumer culture. Known as "the plain people," the Amish until recently lived an existence apart from the mainstream culture, eschewing modern conveniences, living simple, frugal, and uncomplicated lives. But this all changed in at least one community. Like the Amish in other parts of the United States, members of the 20,000-strong Amish community in the vicinity of Topeka, Indiana (120 miles east of Chicago), were slowly assimilating with the general population and its outside economy. The prospect of high wages (and attendant increase in standard of living) in the modular home and recreational vehicle (RV) industries in that part of Indiana was too tempting for many Amish to resist.

By 2007, according to Steven Nolt, a history professor who has studied that particular community, half of the Amish men in northern Indiana were employed in manufacturing, making an average of $30 an hour. This resulted in a marked increase in discretionary income for the Amish, most of whom had worked small farms until this shift. Soon a "keeping up with the Joneses" ethos began to flourish. People wanted bigger weddings, fancier carriages, and expensive horses. Some Amish families bought a second home on the coast of Florida and spent freely on hobbies and family vacations. Instead of getting around by carriage, many took taxis for a day of shopping or a night out at a nice restaurant. The good times were not to last, however. The RV and modular home industries tanked, and along with them, the Amish's newfound consumer lifestyle.

But, as with the rest of us, the current financial sobriety brings to the Amish the potential of a silver lining. As Steve Raber, a thirty-seven-year-old Amish furniture manufacturer, told the *Journal*, "When you have plenty of money, you have a tendency to slowly drift away." Now the economic downturn is encouraging a return drift. The sour economy has spurred a back-to-basics movement in this neck of the Amish world, along with a renewed sense of community. If only it could work such miracles with the rest of us![1]

There is much truth in the axiom "It is easier to avoid temptation than it is to resist it." Environmental programming for consumers rests on that axiom. At its heart, it involves structuring your "environment" (the people, places, and things that make up your everyday life) in such a way as to encourage good or discourage bad behavior. The science behind this technique grew out of careful laboratory research which demonstrates that the likelihood of a response (behavior) is highly influenced by the absence or presence of "environmental cues" associated with that response. A simple example of environmental programming is blocking eBay from your computer or the Home Shopping Network from your television, if these tend to prompt you to overspend. It's simple: if you don't have easy access to these sites, you are much less likely to abuse them.

The environment, as defined above, has a powerful influence on our behavior. We do most things on what we might call "automatic pilot." We hardly have to pay attention to stoplights and intersections when driving to work or to the mall; and when we greet people we already know, we expend little or no cognitive effort. In the long term, the regularity of these types of behaviors may depend upon the consequences they generate (positive consequences will reinforce the behavior), but at any given moment in your day it is the stimuli in your environment that dictate your behavior. The good news is that environmental programming— what I refer to as "environmental tweaking"—has enjoyed considerable success and should be the cornerstone of any behavioral change program. You will be amazed to discover that a little tweaking of your environment can make a world of difference in your spending, your saving, and your attitude.

Tweak (twēk): to make a minor adjustment to.

The suggestions listed in this chapter, culled from the latest research on how to tweak your environment, will help you curtail your spending and actually save some of your hard-earned money. College tuition for the kids gets more expensive every day, and retirement will come sooner than you think. In addition, wouldn't you like to build a rainy-day fund or pay off your mortgage? Let's look at how you can tweak your environment to encourage desired behaviors and achieve financial tranquility.

25 TWEAKS TO FINANCIAL TRANQUILITY

- Consider a "plastectomy."
- Pay yourself first.
- Sign up for SMT.

- Use cash only.

- Build a budget.

- Get it in writing.

- Do it yourself.

- Pay as you go.

- Distract yourself.

- Let cooler heads prevail.

- Remember that there's no "I" in team.

- Pull the plug on TV.

- Just say no to the mall.

- Ban temptation from the house.

- Chart your course.

- Pass the envelope, please.

- Use the buddy system.

- Team up for success.

- Drive savings home.

- Go to "home school."

- Go ahead, commit yourself.

- Rent or borrow, but don't steal.

- Get a job.

- Watch your balance.

- Opt for automatic.

1. CONSIDER A "PLASTECTOMY." Let's face it: Americans are addicted to plastic. Our love of credit cards played a big role in causing our current fiscal debacle. We simply spent more money than we could afford. As was discussed in detail in chapter 6, we typically spend more money when we use credit cards than when we use cash, because cards reduce the "pain of paying" and make it difficult to keep track of our spending. So it's time. Get out a good pair of scissors and cut up every credit card you own. Yep, even the store cards.

I know this sounds draconian (in fact, I can practically hear your anguished outcry), but it needs to be done. While facilitating a series of "Dave Ramsey's Financial Peace" classes at church, I broached the topic of a plastectomy, and I barely survived the insurrection. Cries of "It can't be done," "You can't live in this day and age without credit cards," and "I won't be able to rent a car or shop online" rang out from the forty-plus people in attendance. Once I convinced the incensed throng of disgruntled consumers that all of their usual activities could be done with just a debit card, the shouts for my head were replaced by a general sense of effrontery. I'm not sure how many of those people ultimately cut up their credit cards, but we had at least a smattering who did so throughout the thirteen-week course.

Given the impact that the use of credit cards has on our spending, this is one tweak that is guaranteed to make a real difference—and soon. All it takes is a genuine commitment to breaking the hold the consumer culture has on you and a desire to achieve control over your spending and saving.

2. PAY YOURSELF FIRST. In my opinion, this tweak is the best and easiest way to save money. By speaking to your employer and your bank, you can have money from your paycheck earmarked for savings and automatically deducted before it ever reaches your hot little hands. After a few months you will have adjusted (you can build up gradually) and won't even miss it. A few months ago my wife and I were discussing the need to increase the amount of money in our emergency fund (which should be adequate to cover three to six months' worth of routine expenses). Surprised, Julie reminded me that we had done that six months earlier, through an automatic withdrawal into a savings account set up for that very purpose. We do the same for contributions to our 403(b) retirement account and the kids' college funds. There's no way we would be making such regular contributions to these accounts if we had to write a separate check each month.

3. SIGN UP FOR SMT. Related to the above, the Save More Tomorrow (SMT) program was developed by economist Richard Thaler and several colleagues. This program automatically increases a person's contribution to his or her retirement account in concert with any annual pay raises that exist. Because upticks in saving are timed to coincide with pay raises, participants annually increase their retirement contribution without any decrease in take-home pay.

In Thaler and Cass Sunstein's book *Nudge*, the authors offer a compelling story of the effectiveness that an SMT program can have on savings rates.[2] At the midsized manufacturing firm where SMT was first introduced, employees self-selected into three different groups, the first two of which were *not* involved with SMT.

The first group of employees refused to meet with a financial consultant; they opted to consider and decide their savings on their own. This group's savings rate remained stable at 6 percent over the ensuing three years of the study. A second group of employees met with the financial consultant and agreed to increase their savings rate by 5 percent. At the end of the study, their collective savings rate had topped out at 9 percent—not bad at all.

A third group of employees, however, said that they could not afford the "cut" in pay that increased savings would cause. It was this third group of employees that were offered enrollment in the SMT program. A majority of this group agreed to raise their savings rate by 3 percentage points with each raise. At the start of the SMT program, this third group had the lowest savings rate, at about 3.5 percent of their income. After four pay raises, however, this group's savings rate had nearly quadrupled, to 13.6 percent. The SMT group was so successful that many participants hit the maximum allowable contribution and had to find other places for their increasing store of cash. Thaler and Sunstein note that SMT is now available in thousands of employer plans. My suggestion to you: check to see if your employer offers an SMT plan and sign up immediately if so. Thank you, Richard Thaler and associates!

4. USE CASH ONLY. Commit to paying cash for everything (excluding bills) for two weeks. As you know, the pain of paying is high when you have to count out the green. This eye-opening exercise will help you distinguish between needs and wants and make you more aware of where and when you're spending your money.

5. BUILD A BUDGET. Contrary to popular opinion, budgeting isn't a four-letter word. Although budgeting may not be a lot of fun, it is imperative that you construct a basic budget to track the comings and goings of your money. Easy-to-use budgeting forms can be found all over the Internet. Start by building a simple plan that shows your income and estimated expenses for regular monthly necessities such as heat and food, as well as things like insurance that you may pay only quarterly or once a year. If you aren't sure of the amount in some categories, write down your best guess. With a little digging you can come up with a fairly accurate estimate for most expenses.

Categories like restaurant meals, clothes, and entertainment will be the most difficult at first. But take heart: after about three monthly budgeting sessions, you will have a pretty good feel for where the "fat" can be found in your budget. That budget will then become your blueprint for saving and spending. It will give you the information you need to reduce spending and increase the dollar amount you can put away. I can't tell you the shock on people's faces when they realize just how much they're spending on eating out, say, or on *two* car payments.

Because of the large number of people currently unemployed or underemployed, I have included a "Consuming in a Crisis" monthly cash flow worksheet at the end of this chapter. The worksheet was specifically designed for people coping on reduced incomes but will work for anyone. Many families are slow to adjust their lifestyle after a reduction in household income—waiting on average about six months before making needed changes. This delay is a big mistake. If you find yourself suddenly unemployed, complete the attached worksheet immediately, paying careful attention to where extra income can be found or generated and where expenses can be cut. This worksheet or any other budget will help you make decisions on the best ways to spend your money, help you distinguish between needs and wants, and help you match your spending with your reduced income. Most importantly, a good plan will reduce the stress associated with tightened circumstances and encourage discussion among family members about needed lifestyle changes.

Out of necessity, most people will place a higher priority on fixed expenses such as mortgages, insurance premiums, and car payments. Flexible expenses such as food, utilities, clothing, and entertainment have more wiggle room. Ask yourself how each of these expenses can be reduced. Once the budget has been constructed, make sure that you carefully scrutinize and closely monitor all spending. Before buying anything, ask yourself whether you could go without it, could borrow it, or could buy a cheaper substitute. In the case of large routine expenses such as car payments and auto insurance, ask yourself whether you could sell the car and take public transportation or carpool instead. Drastic times call for drastic measures. The silver lining in all of this may be the lesson learned on the importance of saving for such circumstances.

6. GET IT IN WRITING. Keep a detailed journal of your spending for two weeks. Be sure to write down every expenditure—big or small. This will be helpful in establishing how much you spend on certain activities for your budget and encourage you to be more mindful of your spending. You high-tech types can program your cell phone, laptop, or desktop to remind you to input your expenditures each day and check whether you're sticking to your budget. Your phone or computer could send a warning signal when your budget gets low.

7. DO IT YOURSELF. A sizable amount of money could be freed up for savings if we simply mowed our own lawn and cleaned our own house.

My estimate for a six-month mowing season is about $1,200; for a weekly cleaning service, approximately $4,800 per year. That's $6,000 right there. If you started investing that amount per year in a Roth IRA at a 6 percent rate of return, you would have $139,656 in fifteen years. These may be little tweaks, but they can make a big difference.

8. PAY AS YOU GO.   The idea behind this tweak is to cut out major subscriptions and memberships and instead pay for each product or service as it is desired. This covers routine monthly payments for the HBO movie channel, the gym, your favorite magazines and the daily newspaper, and possibly a book- or chocolate-of-the-month club membership. Many of us don't watch enough movies, read enough articles, or work out often enough to get our money's worth out of these subscriptions and memberships.

If you pay with cash, the cost of such expenses is more salient and you'll end up saving money. At first you might want to combine this tweak with the following envelope strategy (discussed in more detail in tweak 16, below): simply place the money that you'd spend on any given subscription or membership for a typical month into an envelope, and use that cash as your funding source to pay for each as needed. My hunch is that you will have extra money left over in each envelope at the end of the month.

9. DISTRACT YOURSELF.   When you were growing up, you probably heard your grandmother chide someone with these words: "Idle hands are the devil's workshop." Your grandmother was right. If something is bothering you, don't just sit there; do something to take your mind off it. If the after-dinner hours are a "danger zone" for Internet shopping, plan to take a walk, read a good book, or simply get out of the house to take your mind off shopping. Numerous laboratory experiments have established the efficacy of this strategy.

10. LET COOLER HEADS PREVAIL.   Give yourself at least twenty-four hours when making a big purchase (one over $100, say). More often than not, once you've left the store you won't go back. When I worked as a retail salesperson, we were told that once customers left they were gone for good. That proved to be true, too: every once in awhile people would come back, but they were the small minority. Use this to your advantage. Many states have a three-day cooling-off period for any sales transactions conducted in one's home—say, the purchase of a vacuum from a traveling salesperson. Potential customers often feel pressured to buy something that, if allowed further reflection they would not. The three-day cooling-off period—like the twenty-four-hour period I'm suggesting here—allows for such reflection. As will be discussed below, this technique works even better when combined with discussing all major purchases with one's spouse or significant other.

11. REMEMBER THAT THERE'S NO "I" IN TEAM. To stick to a budget, it is imperative that spouses communicate with each other about major purchases—again, over about $100. But beware: "little" expenses like meals out and Starbucks beverages add up quickly. Set a budgeted amount (see tweak 5 above) for such items and use the envelope system (see tweak 16 below) to stick to your budget.

The overarching goal of this book is to help you relate to money and material possessions in a healthier way that will realign your priorities and thus increase your happiness. This will be difficult to accomplish if you and your spouse are not on the same page when it comes to money. As I mentioned earlier, arguments over money are the leading cause of marital distress and dissolution. Couples who have vastly different attitudes about money and material possessions will have a tough time finding the happiness we all desire. And don't think having separate checking accounts is a viable option. That strategy is usually followed so that both spouses can avoid the scrutiny that comes from a joint account, from budgeting, and from open and honest discussions about money and spending.

12. PULL THE PLUG ON TV. This is a simple one: we see commercials, and we want the things depicted. And it's not just the commercials; it's also the product placements and the skewed economic reality portrayed by most shows. TV characters are often shown living beyond their financial means—for example, the roommates on a sitcom that have low-paying jobs yet still manage to have a huge apartment in New York City and dress in the latest fashions. Or the Beverly Hills plastic surgeon who looks all of twenty-nine years of age, is beautiful, wears expensive clothes, and seems to show up for work only on an occasional basis. Remember that research we looked at showing that people who watch a lot of TV want more material possessions and think it takes more money to live at their desired level of income than those of us who watch less TV? Get off the couch and get some exercise, and kill two birds with one stone.

13. JUST SAY NO TO THE MALL. The mall is no place for someone trying to quell materialistic desires and save money. Furthermore, there's no such thing as "window shopping." I'm living proof. The central Texas summers can be hot and unrelenting, so my family and I often walk the mall to avoid the heat. I can count on the fingers of one hand how many times we've left the mall without some unanticipated but suddenly "essential" purchase. Avoiding the mall is the very essence of tweaking— implementing a small change that can make a big difference. Whenever I can, I try to muster support for a walk in the heat instead, or a drive to the local YMCA for an inside walk or workout.

14. BAN TEMPTATION FROM THE HOUSE. This suggestion hits you where you live—literally. Tweak your home environment to help you avoid temptation. This may mean getting rid of catalogs and blocking eBay and other such Internet sites where you shop. Be sure to stop the e-mails from your "favorite eBay vendors" as well, along with recommendations from Amazon.com. I don't have to tell you what to do with "sales alerts" that come from your favorite retailers. Anything that tempts you should be removed from the premises.

15. CHART YOUR COURSE. Festoon your living space with reminders that encourage desired behaviors. How about a simple bar chart that shows your growing retirement account smack dab in the middle of the refrigerator door? Or a daily pop-up screen on your cell phone or home computer that tracks growth in your savings account or diminishing credit card debt? How about inspirational posters? Choose ones that remind you of your goals or inspire you to be a better you. I always threaten my daughters that I'm going to get the 1970s poster of a kitten barely hanging one-pawed from a bar with the caption "Hang in there, baby!" This may be a reference that only my Baby Boomer brethren can appreciate—but it always makes me smile.

16. PASS THE ENVELOPE, PLEASE. This envelope system may seem like a low-tech solution to saving money, but it works. It takes awhile to get used to, but once it's up and running it can work wonders. Based upon your monthly budget, simply earmark an envelope for each of the categories you plan to monitor—say, eating out, entertainment, and groceries. Restrict yourself at first to only two or three spending categories, choosing those that are especially problematic for you or that seem to offer the greatest chance for real savings. Place the budgeted amount in cash for each of these categories in their respective envelopes. Having to pull money out of an envelope and count it out helps you become aware of your spending and keeps you current on how much money you have left for the remainder of the month.[3] If you discover that you've depleted an envelope, it might mean staying in one night instead of going out for dinner, using a coupon or choosing a less expensive restaurant, taking a sack lunch to work, or being more careful when shopping for groceries.

17. USE THE BUDDY SYSTEM. Find a kindred spirit—someone like you who is trying to break the chains of materialism and watching how he or she spends money. This might be a person who can act as a role model or mentor for you, or it might simply be someone on the same path. As we all know, having a buddy makes everything we do a little more enjoyable. Commit to regular contact and be sure to remain positive. Regardless of your financial situation, the standard of living in the United States allows you to remove a lot of fat from your lifestyle and still enjoy many of the finer things in life.

18. TEAM UP FOR SUCCESS. If a single buddy can make a difference, what about a *group* of like-minded individuals? The voluntary simplicity movement, mentioned in the previous chapter, consists of people across the country who have banded together to support each other and share their experience and knowledge about how to live a less cluttered lifestyle. Many such groups act as cooperatives, sharing possessions and maybe even a communal garden.

Debtors Anonymous is a group option if you find yourself buried under debt. Structured like Alcoholics Anonymous, with a formal creed and regular meetings, Debtors Anonymous assigns each member a mentor.

I would be remiss if I didn't mention Dave Ramsey's "Financial Peace" program. I have facilitated the program for three years and could regale you for hours about the fear and sorrow that come when money is your master, but I could also encourage you with countless tales of triumph as people—real people like you—reduce or eliminate debt, save for the future, give to charity, and develop healthy attitudes about money. In the current consumer culture, we need reinforcement such as this to keep the advertising machine at bay.

19. DRIVE SAVINGS HOME. Black Monday: that's what I call the day we bought our new Toyota van (about six years ago). It's not that I don't like cars, but they're a big drain on household finances, especially when seen as status symbols, as they so often are in the United States. A family with *two* new cars can easily spend $1,000 every month on car payments.

Look at big expenditures like cars from an opportunity-cost perspective. Put simply, think what you could do with that money if you didn't already have it budgeted. If that aforementioned two-car-payment family had only one car payment, for example, that would be $500 that could be put to better use. For many of us, that might be all we needed to hit the 15 percent savings benchmark. If you're the typical family earning about

$44,000 per year with a take-home pay of about $2,933 each month (an estimated 80 percent of gross pay), a $500 savings each month would be a 17 percent savings rate.

Trust me: you won't be shunned if you drive a car that's a few years out of date. In addition to our 2005 Toyota van, I drive a 1989 Volvo with 167,000 miles on it. I haven't had a payment on it in years. As I tell my kids, I plan to drive it into retirement and beyond. When we choose to opt out of our consumer culture's message that cars are important status symbols, it can have wonderful financial and psychological benefits.

20. GO TO "HOME SCHOOL." Next to your retirement account, our homes are our biggest financial assets. And, as we saw in earlier chapters, a good part of our current economic troubles can be traced to the mortgage crisis. People bought homes they couldn't afford. Follow a few simple rules of thumbs if you are considering a home purchase: (a) pay as big a down payment as possible—and no pulling money out after a home purchase for other purposes, (b) choose only a fixed-rate loan, (c) opt for a fifteen-year term, and (d) never, I repeat never, take out a home-equity loan. Home equity loans have allowed homeowners to dilute their second-biggest financial asset for transient consumer pleasures.

21. GO AHEAD, COMMIT YOURSELF. Even more important than goals is your commitment to accomplish them. Behavioral intentions—in other words, goals articulated as specific things that you'll do—are far better predictors of behavior than attitudes are. Commit to socking away money for that long-anticipated Europe trip, paying off your credit cards, or establishing a rainy-day fund to cover expenses in the case of illness or job loss. The pain of this current recession would have been greatly reduced if everyone had had six months' worth of expense money in the bank. If you've accomplished all of this already, what about the kids' college fund, paying off your house, or adding to your retirement account? And don't forget charitable giving. When your financial house is in order, you can make a real difference in the lives of others. But it all starts with you and your commitment to saving and to reprioritizing your values.

22. RENT OR BORROW, BUT DON'T STEAL.   We are a people who love to own things—even things that have one-time value. Despite my preaching here, I must plead mea culpa. I have bought far too many books that I could have borrowed from friends or the library. We buy DVDs that most likely will be watched only once; we buy condos (hopefully not time-shares) that we use infrequently; we buy RVs that rarely leave their parking spot alongside the house. We need to break the mind-set that we must *own* things. (By the way, does anyone need a high-pressure washer that's been used only two or three times?)

23. GET A JOB.   This may be advice that you don't want to hear, but you may need to get a second job to dig yourself out from under your credit card or other debt. Use every extra penny you earn to pay off that debt. Trust me: you will think especially hard about using your credit cards foolishly in the future when you have a flashback to your extra job on the late shift at the frozen potato factory or the front desk of the local hotel.

24. WATCH YOUR BALANCE.   Use your checkbook to pay for things, and be sure to balance it regularly. Writing checks, as discussed earlier, requires that you write down the amount you're spending. This "rehearsal" of the amount makes spending more painful. Balancing your checkbook gives you a clear idea of where your money is going and an up-to-date idea of your available assets. I know it takes a little more effort than simply throwing down the plastic, but that's the point: the more mindful you are of your spending, the better you can control it.

25. OPT FOR AUTOMATIC.   Set up your car payment, mortgage payment, utility payment, and other such bills on an automatic payment plan to avoid late fees. I worry about any expenditures that are automatic, though, because if you don't carefully review each bill you're likely to overlook possible errors, and the payments are less salient. Having to write out a check for these things each month—particularly for optional expenses—is a good way to increase the pain of your luxuries. It's your call, but if you're often late, you may want to place these bills on an automatic payment plan.

FIVE STEPS FOR
TWEAKING YOUR ENVIRONMENT

By now you're getting the essence of environmental programming—creating an environment that helps set the stage for successful behavior change. Before we leave this subject, let me share with you five simple steps to follow when altering your environment. These steps encompass and shape the above tweaks.[4]

1. TWEAK YOUR ENVIRONMENT TO DISCOURAGE UNWANTED BEHAVIOR. This requires that you recognize the "cues" that lead to such behavior. What are the people, feelings, places, or things in your environment that cause you to stray?

2. JUST PLAIN AVOID ENVIRONMENTS (PLACES AND SITUATIONS) THAT LEAD TO UNWANTED BEHAVIOR. Don't walk at the mall or surf the Internet if you're trying to watch your spending. If you can't avoid a source of temptation—you work at home, say, and have to use the Internet to check your work e-mail—place signs that encourage good or discourage bad behavior in that "hot spot" (temptation zone). Recognize that most unwanted behaviors are the outcome of a chain of events. To shop at the mall, you need to get dressed, drive to the mall, shop, pick something out, take it to the cash register, and pay for it. Try to recognize what chain of events leads to your particular unwanted behavior, because the earlier you can break that chain, the better.

3. TWEAK YOUR ENVIRONMENT TO ENCOURAGE GOOD BEHAVIOR. Use any of the twenty-five tweaks provided above, preferably several (or even many) in combination, or develop your own.

4. PLAN TO DO SOMETHING ABOUT YOUR GOALS. In any change effort, we have to battle the "status quo bias" (or inertia)—a strong tendency to go along with the existing conditions in our lives. As Thaler and Sunstein note in *Nudge*, the vast majority of people suffer from the status quo bias, and we as a society should learn to harness the power of that inertia. The authors suggest that public policy makers should make preferred policies—in other words, policies that favor the public good—the default option in designing new programs.[5] A person who responded to the program by doing nothing, bowing to inertia, would in effect be "choosing" that preferred policy. You can harness the power of inertia as well, as in the SMT program (tweak 3) designed by Thaler and his colleagues. Even if you're not part of a formal SMT plan, you can commit to keeping only a small portion of all pay raises (with the remainder earmarked for savings) or by making regular increases in your retirement contributions. An implementation schedule—a timeline, if you will—also helps you overcome inertia. For example, I know one family that sets aside one Sunday evening each month to establish their budget for the coming month.

5. KEEP PRACTICING THE DESIRED BEHAVIOR. Pay attention to your environment; never relax your vigilance. In addition to watching for warning signs, be mindful of your plans, schedules, "to do" lists, and anything else you've done to prompt the desired behavior. Don't try a tweak just once and call it good: *keep tweaking*—stay away from temptations, whether they are people or places, not just for a trial week but for life. Every single thing you do to tweak your environment to avoid unwanted behaviors and encourage desired behaviors, and every single time you implement those tweaks, will increase your likelihood of achieving your goals.

We've now programmed your environment to help encourage desirable and discourage undesirable behaviors. Behavioral programming, the subject of the next chapter, involves altering the consequences of your behavior instead of changing the cues that prompt it. Human behavior is greatly influenced by those consequences, and chapter 14 will show you how to use that influence for maximum results.

Consuming in a Crisis
Monthly Cash Flow Worksheet

	PRE-CRISIS CASH FLOW	CRISIS CASH FLOW
A. Monthly Income		
Salary 1 (take-home pay)		
Salary 2		
Other income (alimony, child support, rental or investment income)		
Unemployment benefits		
TOTAL MONTHLY INCOME		
B. Monthly Expenses		
House payment or rent		
Second mortgage		
Utilities:		
Water/sewer		
Garbage		
Heating/cooling		
Home repairs		
Association fees		
Car payment #1		
Car payment #2		
Car maintenance (repairs, oil changes, etc.)		

	PRE-CRISIS CASH FLOW	CRISIS CASH FLOW
Gas		
Groceries		
Cell phone plan		
Home phone		
Cable TV		
Internet access		
Eating out: lunches dinners		
Credit card #1		
Credit card #2		
Additional credit and store cards total		
Personal care (haircuts/styling, manicures, pedicures, massages, etc.)		
House cleaning		
Clothing		
Children-related expenses (lessons, sports, orthodontist, tuition, etc.)		
Day/child care		
Installment loans (TV, boat, etc.)		
Student loans		

	PRE-CRISIS CASH FLOW	CRISIS CASH FLOW
Insurance:		
Dental		
Life		
Car		
Medical expenses (Prescriptions, co-pay, medical supplies, etc.)		
Dry cleaning		
Entertainment (movies, golf, tennis, hobbies)		
Vacations		
Memberships (Health club, country club, etc.)		
Pets (food, vet, boarding, etc.)		
Child support, alimony		
Charitable giving		
Subscriptions (magazines, newspapers, etc.)		
Savings (401k, IRAs, CDs, etc.)		
TOTAL MONTHLY EXPENSES		

C. Monthly Cash Flow* (Income—Expenses)

No doubt your crisis monthly cash flow is in the red (negative number). You have several options. First, you can increase your income by getting a part-time job while looking for employment. If not working, your spouse can find a job. Two, from the long list of expenses, find areas in which you can cut back. Three, you will most likely be able to increase income and drop expenses as well. Any money in savings can be used to stay afloat until you find suitable employment.

— 14 —

THE CARROT
AND THE STICK:
BEHAVIORAL
PROGRAMMING
FOR CONSUMERS

I am indeed a king, because I know how to rule myself.

—PIETRO ARETINO

"Act Your Wage" screamed Dave Ramsey, radio personality and financial guru, from a fourteen-by-forty-eight-foot billboard alongside Interstate 30 in Dallas, Texas. That sentiment about sums it up. We all need to "act our wage." But what's the carrot that encourages us to do so? Two words: financial peace—words that rarely go together in today's consumer culture. A majority of Americans worry about their financial situation and argue with their spouses and other family members over money. Financial peace is that state of mind where you no longer worry about money, no longer go careening from one financial crisis to another. It's a state of serenity and security we experience when money has been put aside for the future and for the inevitable emergencies we will all face. Sadly, however,

70 percent of Americans live paycheck to paycheck. That means they're only one paycheck removed from financial disaster.

But it doesn't have to be like this. As Ramsey puts it, "If normal is broke, be weird." And, as you have seen in our discussion of self-control, this is not rocket science. Ramsey claims that personal finance is 20 percent knowledge and 80 percent behavior.[1] I tend to agree with that breakdown, but I'd replace "behavior" with "willpower" or "self-control." In other words, achieving financial peace is really a question of a little knowledge and a lot of willpower—enough to stay the financial course when all those around you appear to be attending a year-round twenty-four/seven party. The knowledge component of financial peace comes in the form of the previous chapter's suggestions on how to program your environment and this chapter's discussion of behavioral programming. Combine that knowledge with your willpower, and . . . be weird!

As I mentioned earlier, behavioral programming involves altering the consequences of a behavior instead of changing the cues that prompt it. This, then, is the *C* of our **antecedents-behavior-consequences** (ABC) model. In this chapter we will look at ways that we can assign positive consequences (self-rewards) to those behaviors we want to reinforce and negative consequences (self-punishments) for those we want to discourage. All such consequences are either internal or external. For example, self-praise, self-criticism, and pleasant or unpleasant mental images are internal consequences, while tangible rewards for good behavior (new clothes, an extra nap, time with friends) or the removal of such things when you don't keep up your end of the bargain are external consequences.

THE CARROT: USING SELF-REWARD
TO CHANGE BEHAVIOR

As anyone who has seen her credit rating restored through diligence after an early financial fiasco knows, human behavior is influenced by its consequences. When a behavior is promptly followed by an effective "reinforcer"—anything that encourages the behavior it follows—that behavior is more likely to happen again. We call this the "law of effect."[2] There are two types of reinforcers. Positive reinforcement strengthens a response by rewarding a particular behavior with something you enjoy— for example, watching a movie or eating a treat after you finish your chores. Negative reinforcement strengthens a behavior by reducing or removing an aversive stimulus (something you don't like)—for example, ceasing to spend money foolishly to avoid your spouse's nagging, balancing your checkbook so that you can stop worrying about whether you have enough money for groceries, or opening an IRA in order that your worried parents can rest a bit easier.

The first task in any self-reward project is planning the circumstances when specific rewards will be provided. The father of behavior modification himself, twentieth-century psychologist B. F. Skinner, is quoted as saying, "The way positive reinforcement is carried out is more important than the amount."[3] To maximize the success of your self-reward program, you must decide the following:

1. What behavior to target

2. Which rewards to use

3. When and how to administer the rewards

4. How to ensure that your self-change will be maintained

Over a century of research in the field of psychology provides broad support for the use of rewards to help elicit behavior change. The good news is that it works even when *self*-administered. Research has shown that it is more effective than no treatment and *just as effective* in most cases as going to a therapist. Psychologist and author Clayton E. Tucker-Ladd found that even therapists themselves feel that self-help is effective, and most recommend various self-help techniques to their clients.[4]

The saying "Different strokes for different folks" summarizes well the variety of self-rewards that can be used to encourage a desired behavior. A list of possible reinforcement categories follows, but you know best what will work for you.[5] Choose one or two rewards from the list, or come up with your own.

PHYSICAL PLEASURES. Rewards of a physical nature can be foods—a can of soda, a snack or other small treat, a fancy dessert, or a romantic candlelight dinner—or delights such as a nap, a backrub, a weekend with your spouse (without kids!), or a full-body massage. (If you're prone to weight gain, be careful about using foods as treats.)

SOLITARY ACTIVITIES. Allow yourself the time and freedom to daydream, enjoy the sunset, people-watch, read an article from the newspaper, play with your pet, read a book, relax, take a walk or a bike-ride, start a new hobby, complete a do-it-yourself project, paint a picture, or write a story.

SOCIAL ACTIVITIES. These can range from talking to a friend on the phone, having someone over to your home, planning a neighborhood get-together, helping a neighbor prepare for a vacation, joining a club, going dancing, or doing volunteer work (a great way to feel good about yourself).

SELF-APPRECIATION AND PRAISE. Affirm yourself whenever possible by saying to yourself, "Good job!" or "You made it!" or by sharing your success with a good friend or family member; take a minute to enjoy what you have accomplished in life and congratulate yourself for being who you are.

REGULAR, EVERYDAY BEHAVIORS. Even little things can be effective reinforcers. Stretch your legs only after you've finished paying the bills, take a short walk every two hours at work, fill up your coffee mug after a good deed, write an e-mail to a friend after you've caught up on must-do correspondence, read a couple of cartoons from a collection of your favorites after you've done the dishes, or make a quick visit to a fun website after a certain task has been completed.

MATERIAL OBJECTS OR MONEY.   Material reinforcement can range from inexpensive knickknacks, to more expensive items such as sunglasses, books, clothes, a new phone, a better computer, updated furniture, or cold, hard cash. There's nothing wrong with a little splurging *after* you have accomplished your goals. Remember: the point is not to be a spend-thrift or a miser, but to fall somewhere between these two polar oppo-sites. Material objects can work, but they take a bit more effort in their execution than does, say, simple cash or the gift of time for relaxation. A creative variation is to give up some of your treasured possessions and earn them back as you attain your self-help goals.

To be effective, a reinforcement system has to be simple and easy to administer. The ultimate goal is to shift from regular rewarding of the behavior, to intermittent reinforcing, to eventually finding enjoyment in the behavior itself (intrinsic satisfaction). At first, however, it's best to re-inforce the desired behavior every time it occurs. Under such conditions, learning occurs rapidly.

An effective reinforcement system also requires regular self-monitoring of the sort we talked about in chapter 12. To ensure that you maintain the desired behaviors that you're reinforcing, you have to monitor the target behavior (for example, spending rate, savings rate, or credit card usage) at least every week, if not daily for the first three or four months. Be forewarned, however: for crucial financial behaviors, you will prob-ably need to monitor your behavior at regular intervals for the rest of your life. If you don't believe me, think of all the people you know who have backslid on their budget or built up their credit card balance again after swearing never to use credit cards again (or even after cutting their cards up). I worked in the consumer loan division of a major bank in a past life, and I can't tell you how many times people took out con-solidation loans to pay off their credit card debt, only to come in a year later in worse shape than before their debt consolidation. While moni-toring your behavior is a lifelong commitment, it's not an onerous bur-den when done faithfully. That's because careful monitoring will enable you to catch small problems before they become big ones. Caught early (before momentum can build), small transgressions are much easier to correct.

Commitment Contracts

A good way to ensure that your self-reward program is executed properly is to put it in writing—that is, to make a contract with yourself (often termed a "self-contract," a "commitment contract," or a "social contract" if you get others involved). Be precise as to the target behavior and the consequences of keeping or breaking the contract. The higher the consequences (stakes), the more likely it is that you will reach your goal. I strongly suggest that you involve others in that contract's administration. If you do so, make sure that those people know what behavior to look for and what consequences will result. Many people find that it works best to use a combination of self-reward and self-punishment. At the end of this discussion I present an example of a self-contract and list the characteristics that make such a contract effective. Use these tools as a starting point, but don't hesitate to tailor your contract to your particular needs.

A high-tech version of a commitment contract can be found online at stickK.com, a website cofounded by Dean Karlan, an economics professor at Yale University. His research has shown that using commitment contracts can "more than triple your chances of success" (from the website, which also explains that "the [second] letter 'K' [in the name] is the shorthand symbol for 'contract' used in legal writing"). The commitment contract concept is based on two well-known principles of behavioral economics: (1) people don't always do what that say they want to (no big surprise there), and (2) incentives work in getting people to achieve their goals. StickK addresses these principles by establishing incentives and assigning accountability.

A large banner ad (over a backdrop of dollar bills) on the stickK website proclaims that you "will increase your odds of success by putting money on the line." Money—now that's clearly an incentive. StickK takes your credit card information at the time you set your goals. If you don't make each incremental goal—say, saving $200 each week—stickK gives a mutually agreed-upon donation to a charity. Worse yet, you can opt for the money to go to an "anti-charity"—that is, an organization whose views you oppose. For example, stickK might make a donation to the National Democratic Party on your behalf when you're a died-in-the-wool Republican, or to the National Rifle Association when you're in favor of tighter restrictions on guns. The anti-charity idea appears to be an effective tactic, with stickK-reported success rates as high as 85 percent. Accountability comes in the form of friends and family you recruit, and others from the website who volunteer, to keep you honest and help you make progress toward your goals. (Another website, www.successorelse .com, uses only punishments to encourage people to achieve their goals, but otherwise it operates very much as stickK does.)

StickK works, or can work, for several reasons. First, whereas in the case of a traditional social contract with a friend or family member who may not hold you accountable, that's no problem for the folks at stickK. Second, with the online option you can hold yourself accountable to a larger circle of friends and family members as well as to the entire online stickK community. Third, stickK sends e-mail reminders, provides a commitment journal to chart your progress, offers expert advice in many cases, and requires an end date—an essential of any good goal. Finally, best of all, it's free—unless, of course, you don't keep your end of the bargain. I encourage you, as the folks at stickK do, to "put a contract out on yourself."

You don't have to go the high-tech online route, though. You can easily draft your own document using the following guidelines, condensed and adapted from a list compiled by Clayton Tucker-Ladd:

- Conditions of contract must be clear.

- Contract should stress the positive, but punishments can be used as well.

- Other people should be included in the contract, if possible. (Be sure to thank your support team often for their encouragement along the way.)

- Contract should include provisions to monitor progress and make adjustments when needed.

- Goals specified should be "SMART"—specific, measurable, achievable, realistic, and timed.

- Contract should state specific consequences—rewards and punishments that will be motivating for you. Rewards need not be big; punishments need not be harsh.

- Contract should specify the timeframe for assigning rewards and/or punishments, with immediate consequences initially, then increasingly spaced out as the target behavior becomes the norm.[6]

Although putting your proposed goals and consequences in writing highlights the specifics of your change program and helps keep you on task, you need to remain flexible: if you fall behind, give yourself a break; if you make better progress than expected, up the reward.

Sample Self-Contract or Social Contract

DATE: _____

SELF: _____

OTHERS (THIS WILL BE RELEVANT ONLY IF YOU'RE MAKING A *SOCIAL* RATHER THAN A *SELF*-CONTRACT): _____

REFEREE: _____

GOAL: _____

Agreement between Parties

Self:

Others:

Consequences (Stakes)

Provided by self
(if contract is kept):

(if contract is broken):

Provided by others
(if contract is kept):

(if contract is broken):

Review schedule (specify intervals when progress reports will be taken and rewards given for progress):

CONTRACT COMPLETION DATE: _____

YOUR SIGNATURE: _____

WITNESS: _____

Source: Adapted from Michael J. Mahoney and Carl E. Thoresen, Self Control: Power to the Person, *(Monterey, CA: Brooks/Cole Publishing, 1974), 51.*

SUMMARY: THE FIVE STEPS
OF SELF-REWARD

I'll focus here on positive reinforcement, because it's something we all administer often (even in the absence of any formal change effort) and understand easily: do a "good deed" and get a reward—a night out if you open an IRA, or a family vacation if you stay on budget. What follows next are the primary steps involved in using positive reinforcement to increase desirable behaviors. You'll hear echoes of the earlier discussion of the commitment contract, but these steps are essential to *any* change effort, even in the absence of a contract.

1. IDENTIFY CLEARLY THE DESIRED BEHAVIOR.   Be specific as to what the desired behavior entails (contributing regularly to your IRA, saving money for an emergency fund, developing a household budget, etc.). Set subgoals—the little steps required to get you to your desired behavior— as well as your ultimate big goal. Be sure to reward these baby steps along the way. Make sure that all your goals are "SMART"—specific, measurable, achievable, realistic, and timed.

2. CHOOSE APPROPRIATE REWARDS (REINFORCERS).   There are many different types of rewards available, as we saw earlier. Let your imagination supplement the list of reinforcers given above. Be creative! Choose rewards that are available, are under your control, and are powerful enough to motivate you to break old habits and establish new, healthier responses. Don't give so many rewards that they lose their power, don't give rewards for easy tasks, and don't reward tasks you already like. Increase rewards as tasks get more difficult, and consider phasing out tangible rewards and moving to positive thoughts and images as the behavior becomes ingrained. Behavior that is satisfying in itself is the most likely to continue.

3. PLAN YOUR REINFORCEMENT SCHEDULE.   At least initially, rewards should be immediate and frequent, granted every time the desired behavior occurs. As the behavior becomes more common, you may choose to reward yourself less frequently and may want to switch from tangible rewards to mental rewards such as pleasant thoughts. A written commitment contract, discussed earlier, is particularly helpful at establishing your reinforcement schedule.

4. GET STARTED.   Now it's time to put your plan into action; if you've opted for a written document, it's time to carry out the contract. You can adjust your plan or contract later as needed. Don't worry about perfection—just get started! However, be flexible if you see that your goals are too high or low, or that certain reinforcers are not working.

5. BE VIGILANT AGAINST BACKSLIDING.   Continue to closely monitor your behavior even after you have achieved your ultimate goal. As was noted earlier, it is much easier to reverse an unwanted behavior when it is detected early. As with drinking and eating, keeping materialistic consumption under control is a lifelong battle. Not a day goes by that we don't hear the current consumer culture extolling virtues that are antithetical to true happiness. We live in a place and time where the majority of people choose to follow the shiny-objects ethos and worship at the altar of consumerism, and we have to consciously choose a different path. Psychiatrist Viktor Frankl, Holocaust survivor and author of *Man's Search for Meaning*, may have put it best when he said, "Between stimulus and response there is a space. In that space is our power to choose our response. In our response lies our growth and our freedom."

THE STICK: USING SELF-PUNISHMENT TO CHANGE BEHAVIOR

The desired effect of self-punishment is the opposite of self-reward: where self-reward increases a behavior, punishment decreases it. For our purposes here, punishment is any consequence that decreases the likelihood of the preceding behavior occurring again. As with self-reward, self-punishment can be either positive or negative. Positive punishment involves delivering an undesirable consequence; negative punishment involves withdrawing a desirable consequence. Examples of positive punishment include conjuring up an image of going to bankruptcy court if you continue to misuse credit cards or assigning yourself extra housekeeping chores if you don't stick to your budget. Examples of negative punishment include taking away TV privileges or lunch out with friends if you go on a shopping spree.

Like reinforcers, punishments can take on many forms depending on the individual. The list below shows five categories of potential self-punishment.[7] These are only suggestions, however. Anything that you wish to avoid is a good punishment *for you*. Be creative!

1. PHYSICAL DISCOMFORTS. Try delaying a meal, doing extra exercise, taking on a chore you don't like, or taking *away* something physically pleasurable that you enjoy—no dessert, say, or no nap.

2. RULES, PENALTIES, AND CORRECTIONS. Establish rules such as "You can't buy a big-screen TV until you've funded your IRA for the year." (These can also be seen as delayed rewards.) Set penalties in the form of unpleasant tasks required if unwanted behavior occurs (for example, having to wash the kitchen floor if you fail to write out the bills on schedule). Call for "corrections" by which you make up for the harm done by an unwanted behavior (or "overcorrections" where you have to do more than just make things even—you have to do extra chores, say, for chores missed).

3. SELF-CRITICISM. This is something we all do rather well, and it can be an effective punishment. Be sure to not be overly critical, however, and end your criticism with an affirmation such as, "I know I can do better than this." Face up to the possible consequences of your behavior, perhaps by making a list of the negative consequences of your spending or your materialistic values. Simply ruminating on such negative outcomes can be an effective punishment, especially for highly self-critical people.

4. CONJURE MENTAL IMAGES OF THE NEGATIVE CONSEQUENCES OF YOUR BEHAVIOR. Ask yourself, "What will my marriage be like if I continue to overspend?" Or how about this? The next time you're considering buying something, picture that object sitting on the table at your next yard sale with a $1 price tag—right next to that macramé beer can "hat" (a 1970s classic) and all of the other similarly priced junk you've accumulated over the past year. That's a sure way to dampen your consumer longings! A visit to Debtors Anonymous might be helpful in conjuring up additional possible negative outcomes of overspending.

5. INVOLVE OTHERS IN YOUR PUNISHMENT. As discussed earlier, self-contracts are more effective when others are involved in administering punishments as well as praise. From personal experience, I can say that a public proclamation to family members, colleagues, and the local newspaper of my intent to write this book was a major motivating factor behind its completion. Likewise, when others know that you're trying to save money, you're a bit more likely to stick to your budget.

Again these are only suggestions. Anything that you wish to avoid is a good punishment. Be creative.

SUMMARY: THE FOUR STEPS OF SELF-PUNISHMENT

If you feel the need for change and you think that you'd be well motivated by self-punishment, how do you proceed? Here are the four major steps involved in using self-punishment to decrease undesirable behaviors.[8]

1. IDENTIFY THE UNWANTED BEHAVIORS.   Be specific. What behavior do you want to reduce or eliminate, and where and when does it take place? For example, perhaps you routinely spend more money than you make. (As I have heard it said, "There's too much month/week left at the end of the paycheck.") If that's your big issue, relevant unwanted behaviors could include: (1) eating out, (2) paying for purchases with credit cards, and (3) asking others for loans. That's where you'll want to focus your punishment. (Before using punishment, however, take a close look at the environment in which the unwanted behaviors occur. As discussed in chapter 13, if you can tweak your environment to decrease the odds that your unwanted behaviors will occur, punishment may not be needed.)

2. DECIDE ON APPROPRIATE PUNISHMENTS.   Browse through the list of punishment categories given above in search of actions that would be effective motivators in your specific case. Let your imagination supplement that list with punishments of your own devising.

3. MAKE UP A SELF-CONTRACT FOR THE DETAILS OF THE PUNISHMENTS. Your contract should look very much like the sample contract provided earlier in this chapter. The contract will probably contain both reinforcers and punishments. As we discussed earlier, you can also go to stickK. com or similar websites and spell out the details of your commitment contract. Be sure to enlist as many supporters as possible.

4. BUILD IN BACKUP FOR THE UNWANTED BEHAVIORS. Remember that punishments simply stop unwanted behaviors. You'll want to encourage alternate, desired behaviors as well—and practice those desired behaviors before starting your punishment schedule. Again, it's not enough simply to stop an unwanted behavior; you must replace it with a desired behavior.

We live in a materialistic world. That's a given. So what are you going to do about it? Chapter 15 reveals that the hope for a better world begins with you. It really is your money or your life. You can't have it both ways.

YOUR MONEY
OR YOUR LIFE

It is the preoccupation with possessions, more than anything else, that prevents us from living freely and nobly.

—HENRY DAVID THOREAU

Most life goals can be placed in one of two opposing camps: personal happiness or service to others. In other words, goals are typically *self*-oriented or *other*-oriented. The problem is that pursuing one type of goal comes at the expense of goals in the other camp. So who is the happiest— a person who devotes his time to the pursuit of personal happiness or the person who devotes his life to helping others? The following brief exercise may help you fashion an answer to that question, which is one of life's most fundamental.

First, list ten people whom you know well. Next, rate each person as either happy or unhappy. Finally, rate each person as either self-centered or other-oriented. Do you see a pattern? Researchers have: they've found that happy people are ten times more likely to be other-oriented than self-centered. This suggests that happiness is a by-product of helping others rather than the result of its pursuit.[1] Rabbi Harold Kushner, author of

When Bad Things Happen to Good People and *When All You've Ever Wanted Isn't Enough*, may have put the dilemma best when he said:

> Happiness is a butterfly—the more you chase it, the more it flies away from you and hides. But stop chasing it, put away your net and busy yourself with other, more productive things than the pursuit of personal happiness, and it will sneak up on you from behind and perch on your shoulder.[2]

The pursuit of power, personal glory, money, and material possessions—all important components of the American Dream—has undermined our happiness. Such pursuit comes at the expense of personal growth, good relationships, and service to the community. It's time that each of us decides whether our primary life goal will be the pursuit of personal happiness or service to others. Your choice will have a profound impact on your life. One thing's for sure: doing great good is unlikely if your primary focus is seeking personal happiness.

We all want to live a life that matters. In his book *When All You've Ever Wanted Isn't Enough*, Kushner tells a story about an old man who went for a stroll in the woods and got lost. The man wandered for hours trying to find his way back to town, going down one path and then another—but none led to town. As he was trying yet another path, he bumped into a fellow walker and exclaimed, "Thank God for another human being. Can you show me the way back to town?" The other man replied, "No, I too am lost. But we can help each other in this way. We can share with each other which paths we have taken and have been disappointed in. This will help us find the path that leads to town."[3]

This story reminds me of why I wrote this book. Many of us have been on the path of materialism, and we've learned that it doesn't lead to "town." As detailed in the preceding chapters, happiness cannot be purchased. If *I* haven't convinced you of that yet, listen to what funny man and actor Jim Carrey had to say about fame and fortune: "I think everybody should get rich and famous and do everything they ever dreamed of so they can see that it's not the answer."[4] It is not the goal of this book to change the world—that would be foolhardy and futile. And yet, although nothing around you will change as a result of your reading these pages, something inside you might shift. Your world can be a different place when you shed the blinders of materialism and find your authentic self, not some one-size-fits-all prepackaged self sold by Madison Avenue.

Aristotle once said, "Happiness is the meaning and purpose of life, the whole aim and end of human existence."[5] Our collective attempt to find our happiness in material possessions has fallen short of its goal. Fourth century teacher, philosopher and priest St. Augustine may have put it more eloquently when he said, "My soul was sick and covered in sores, and it rubbed up against material things in a desperate attempt to relieve the itching. But since material things have no soul, they cannot be loved."[6] The primary goal of this book is to make the argument that lasting happiness lies outside of the consumer realm, beyond the shiny-objects ethos. I have attempted to achieve that lofty goal by putting to work all I know about persuasion and attitude change. Marketers, as a lot, know more than a little about such things—after all, it's how we make a living.

ATTITUDES PRECEDE BEHAVIOR

Attitudes are learned predispositions—amalgamations of our many and varied life experiences. They are important because they both dictate and help us understand our behavior. In most instances, a negative attitude leads us to avoid an "attitude object" (whether a person or activity or event), while a positive attitude makes it more likely that we will be favorably predisposed toward that object. Attitudes consist of three components: cognitive ("thinking"), affective ("feeling"), and conative ("doing"):

- The cognitive component of an attitude is what we know (or *think* we know) about something. Our beliefs can be based on knowledge (first- or second-hand experience), opinions, or values we hold.

- The affective component of an attitude is our feelings of like or dislike, based upon our reaction to our cognition regarding the attitude object—say, materialism. This component is what most of us think about when someone refers to a person's attitude.

- The conative component of an attitude represents our behavioral intentions in regard to an attitude object. Someone who has a materialistic orientation would have behavioral intentions consistent with such values—for example, buying things to impress others.

Advertisers have two basic paths when attempting to mold your attitudes and shape your subsequent behavior. The first path is what marketers call the "think-feel-do model" of attitude change. In this model, we appeal first to your head; our primary objective initially is to teach or share with the audience information about the attitude object. The "feel" stage, which comes next, is impacted by the information that the consumer retained in the "think" stage—that is, any positive or negative feeling you have about the attitude object is based upon your reaction to the knowledge, opinions, and values acquired in the "think" stage. These positive or negative feelings then impact your behavioral intentions in the conative stage of the think-feel-do attitude model. Positive feelings increase the likelihood of buying a particular product or accepting a particular argument.

Think about that model in terms of this book. The first eleven chapters were an attempt to inform and persuade you about materialism and its impact on your well-being. Much of the information shared and arguments made may have been new to you, or may have been in conflict with how you initially felt about materialism. If I was persuasive enough in those chapters, you may now have more positive feelings toward a less materialistic lifestyle. These newfound feelings make it more likely that you will change how you live your life. A less materialistic outlook will affect not only how you spend and manage your money, but also how you live your life in a broader sense. The time, money, and energy freed up by a less materialistic lifestyle open up a world of possibilities including (but not limited to) self-enrichment, more fulfilling personal relationships, and heightened involvement in community activities—in sum, a more meaningful life.

Remember that I said the think-feel-do model was one of *two* approaches to attitude change? What about the second path? In this approach marketers reverse the order of the three components into the do-feel-think model. Instead of appealing to the cognitive component of your attitude first, this model initially tries to get you to act. Now think of this book again. Chapters 12 through 14 on self-control used the do-feel-think approach. If I can get you to try something, that new action will lead to positive feelings, which in turn will impact the formation of the cognitive component of your attitude. Much research tells us that attitudes based on firsthand experience—in other words, things you've tried yourself—are much stronger and better predictors of behavior than attitudes formed from secondary sources of information like advertising or word of mouth. It is my fervent desire that when you try environmental and behavioral programming to improve your self-control, those efforts will further reinforce the attitudes shaped in earlier chapters through my think-feel-do efforts.

LIVING A LIFE OF MEANING

We have been struggling with how we relate to money for centuries—no, millennia. Over 2,000 years ago Aristotle had this to say about money and materialism:[7]

> There is no sense in producing or acquiring more shoes than can possibly be worn. This is self-evident. With regard to money, however, which has become exchangeable against everything, the illusion arises that it is good to accumulate it without limit. By doing so, man harms both the community and himself because, concentrating on such a narrow aim, he deprives his soul and spirit of larger and more rewarding experiences.

Clearly, the roots of our passion for material possessions run deep. What's different today compared to earlier times is that our obsession with shiny objects has undermined the very thing we've been hoping to achieve with all this consumption—our happiness. Our sense of self-worth, our relationships, our community involvement, and our financial well-being have all suffered for our efforts. Money and material possessions have become an end in themselves, not a means to an end.

Despite purchasing, consuming, and amassing more and more products year in and year out, we are no happier today than we were five or ten or forty years ago. Some would say we are even less happy, experiencing higher incidences of mental illness than past generations. My hope is that, if we take some time to consider our current dilemma, we may see a glimmer of sunlight in what is otherwise a fairly gloomy forecast. But the transition from a life of materialism to a life of meaning—the life illuminated by that glimmer of sunlight—will not be an easy one. Materialism is so deeply embedded in our consumer culture that it will take a concerted effort to step away from the shopping cart, so to speak.

But it's a struggle well worth the effort. As the previous pages attest, materialism negatively impacts not only each of us individually, but those around us as well, along with the larger communities in which we live. Changing ourselves is tough enough, but the world as well? Hear what a wise person once said about the power of individuals to change the world:[8]

> God said to me: Your task is to build a better world. I answered: How can I do that? The world is such a large, vast place, so complicated now, and I am so small and useless. There's nothing I can do. But God in his great wisdom said: Just build a better you.
>
> —*Anonymous*

THIS BOOK'S APPROACH IS VERY MUCH IN TUNE WITH THE ABOVE SENTIMENT. It's up to you to begin the process of change. I acknowledge that it won't be easy. Think of a salmon swimming upstream to spawn: anyone who eschews materialism will be swimming upstream in a society convinced that material possessions are the road to happiness.

But the good news is that, even in our current consumer culture, we are not doomed to a life focused solely on the pursuit of wealth and material possessions. Many of us are frustrated with the myopic pursuit of money and the trappings of wealth and are searching for a simpler way of life. Many of us are seeking out voluntary simplicity, choosing to limit our material consumption in order to free our resources of time and money—in short, seeking satisfaction through nonmaterial aspects of life.

Living a simpler lifestyle does not mean selling all of your belongings and moving to a kibbutz. It's about balance and moderation. Financial aspirations are not in and of themselves bad; it's when they are disproportionately favored over more intrinsically satisfying goals that they become toxic. Living a simpler life just means that each of us carves out the time and energy to grow as a person, have meaningful relationships with family and friends, and become involved in something bigger than ourselves.

We are a nation adrift, having lost sight of the true American Dream. We are a people afflicted with the diseases of boredom and meaninglessness. We have lost the purpose in our lives. A wise man once said, "To live a meaningful life, you must live a life of meaning."[9] If I were to ask what's more important in life, making and spending money or your family, most people would answer "family." But our behavior, as a people, isn't consistent with that attitude, as we have seen. Americans work more hours on average than people in any other industrialized nation. We work to earn money to spend and then work more to pay for our spending, sacrificing time with our families on the altar of stuff.

It was Oscar Wilde who wrote, "In this world there are only two tragedies. One is not getting what one wants, and the other is getting it." Wilde was attempting to warn us that no matter how successful we are and how many possessions we accumulate, we will not be satisfied. Our souls hunger instead for meaning—a hunger that money and material possessions cannot satisfy. For something to be considered meaningful and fulfilling, it must pass what I call the "deathbed test." If asked on your deathbed to complete the following sentence, "I wish I had spent more time_____," how would you respond? I doubt that many of us would wish we'd spent more time at work or shopping. No, I'm guessing that a lot of your responses would revolve around spending more time with family, in church, with friends, or helping others.

In the past I've encouraged my students to draft their obituary as part of an extra-credit exercise, asking them to write it as they wished it to appear after a long life. I know it sounds morbid, but this assignment is an excellent way to expose the gap between how people live their lives and how they want to be remembered. It's a sobering experience at any age. The real benefit, of course, is that the writer gets to confront any such inconsistencies before it's too late. I can't tell you how many students have told me that this experience was a real wake-up call for them—one that I hope has changed many lives. We all must ask ourselves, if our obituaries were written today, would we be happy with what was written? It was the Greek philosopher Socrates who said, "An unexamined life is not worth living."

Our need to live a life of meaning is neither a physical nor a psychological drive. It is a spiritual need—a need to be part of something that transcends the everyday and connects us to something bigger than ourselves.[10] The Golden Rule, which is an important part of many religions, has been given that name because it is the highest rule of life. The secular wording of the Golden Rule would go something like this: "Do unto others as you would have them do unto you." In the New Testament Sermon on the Mount Jesus says, "Therefore all things whatsoever ye would that men should do to you, do ye even so to them" (Matthew 7:12). The Old Testament version in Leviticus reads, "Thou shalt love they neighbor as thyself" (19:18). The Islamic Golden Rule reads, "No one of you is a believer until he loves for his brother what he loves for himself" (Al-Nawawai, 13).

The Golden Rule is a real acid test for those who claim a spiritual connection and/or attend church. Psychologist Clayton Tucker-Ladd has asked hundreds of college students the question below, which I would now like to ask you:[11]

Is it morally just and fair to be free to have plenty to eat,
nice clothes, luxuries, time and money for fun, TV, and
comforts, while others in the world are starving, uneducated,
and in poor health?

CIRCLE ONE: Yes   No

As you think about your response, remember back to the survey we discussed in chapter 7 which found that freshman entering college are increasingly placing a much higher importance on being well-off financially over developing a meaningful philosophy of life; likewise, more college students over the years have answered yes to the above question. My intention isn't to point a finger at college students. Even a cursory glance at the majority of our lives would reveal that the Golden Rule has taken a backseat to our desire for material possessions. Clearly, we have lost our moral compass.

MORALLY SPEAKING

Morals help us to distinguish between right and wrong. According to Tucker-Ladd, it is critical that we take a careful look at our morals for a number of reasons: (1) morals can guide our lives in the pursuit of noble goals and keep us from self-serving endeavors, (2) morals can inspire and motivate us to live a meaningful existence, (3) a failure to live up to our morals can be the catalyst to improving ourselves, (4) living by our morals can lead to high self-esteem, and (5) in Tucker's words, "professed but unused morals or values are worthless or, even worse, can be construed as phony goodness and therefore used as rationalizations for not changing."[12] That fifth point is particularly relevant to the present conversation because, as I noted earlier, many of us profess that our families are more important than money but behave in ways that are not consistent with that sentiment.

Building a better world begins with each of us individually, and that rebuilding begins with finding our lost moral compass. Developmental psychologist and moral philosopher Lawrence Kohlberg identified a process of moral development that people move through in life (some farther and faster than others)—six stages that gradually lead us toward moral maturity. Perhaps a better understanding of those normal stages of development would help us develop or improve upon our own morals and values.

Before I describe Kohlberg's stages of moral development, consider the following moral dilemma, which (with other dilemmas) Kohlberg posed to children, adolescents, and adults to identify their stage of moral reasoning:

The Heinz Dilemma

In Europe, a woman was near death from a very bad disease, a special kind of cancer. There was one drug that the doctors thought might save her. It was a form of radium that a druggist in the same town had recently discovered. The drug was expensive to make, but the druggist was charging 10 times what the drug cost him to make. He paid $200 for the radium and charged $2,000 for a small dose of the drug. The sick woman's husband, Heinz, went to everyone he knew to borrow the money, but he could get together only about $1,000, which was half of what it cost. He told the druggist that his wife was dying and asked him to sell it cheaper or let him pay later. But the druggist said, "No, I discovered the drug and I'm going to make money from it." Heinz got desperate and broke into the man's store to steal the drug for his wife.[13]

So what do you think, should Heinz have stolen the drug? More important, why was what Heinz did right or wrong? Kohlberg was not interested in whether respondents judged Heinz's behavior as right or wrong (either could be justified); rather, it was the rationale behind their decision that provided insight into their level of moral reasoning.

Kohlberg argued that we progress to a higher level of moral reasoning with each subsequent stage. His six stages can be grouped into three levels consisting of two stages each: the levels are *preconventional, conventional,* and *postconventional*.[14] Only a minority of people progress to the postconventional level—stages 5 and 6—with the rest of us mired in stages 3 and 4 of the conventional level or even stages 1 and 2 of the preconventional level. As our moral reasoning matures (and we move up through the stages), our behavior becomes less selfish and more other-oriented. Where would you guess that *you* fall along Kohlberg's moral reasoning continuum?

PRECONVENTIONAL MORALITY. The preconventional level of morality, the lowest level of moral reasoning, is primarily populated by children, although an increasing number of adults could be classified at this level.

People at stage 1 of this level make decisions based upon the possibility of punishment: a behavior is considered wrong only if someone is actually punished for it. Behavior at this stage is entirely egocentric, with little or no thought given to the possibility that others may have differing opinions. Although Kohlberg designates stage 1 as typical for children between one and five years of age, I know a lot of adults who have pitched their tent here.

Stage 2 can be characterized as "looking out for number one." Behavior is deemed appropriate if it serves the individual's best interest. This stage of moral development is typical of children between the ages of five and ten. Denizens of this stage act in a self-serving manner and have little respect for the rights of others. (Reminds me of waiting in the carpool line at my daughters' grade school. Unfortunately, it's the parents behind the wheel I'm picturing at this stage of development.)

CONVENTIONAL MORALITY. The next level of moral development, the conventional level, includes stages 3 and 4 of Kohlberg's hierarchy. Hold the elevator; many of us will be getting off here if we haven't already departed.

At stage 3 we begin to fill social roles. We desire approval and want to avoid disapproval from significant others—friends, parents (enjoy it while you can), teachers, coaches, and others. People at this stage no longer simply want to please themselves; they look for others' favor as well, and will conform to others' expectations to gain that favor. Conformity is the key. At this stage improper behavior is commonly justified by saying, "Everyone is doing it."

Stage 4 is characterized by law-and-order thinking. People at this stage have internalized society's rules about what is proper behavior and feel obligated to conform. Top priority is given to maintaining social order but with little thought given to the ethical underpinnings of such social rules. Most active members of society remain at stage 4 or lower, where morality is largely dictated by an outside force.

Given that this is the level of majority, it is easy to see how materialism has flourished in our time. Selfishness thrives in the earliest stages of moral development, and an egocentric nature is the perfect breeding ground for materialistic behavior. The battle cry of stage 2, "Look out for number one," typifies the self-focus of materialism. Conformity, an important driver of moral reasoning at levels 3 and 4, also plays an important role in materialism. Those high in materialism are especially sensitive to the opinions of others and place a great deal of importance on their place in the social hierarchy.

POSTCONVENTIONAL MORALITY. The postconventional level of morality, the highest level of moral reasoning, is sparsely populated.

Stage 5 of the postconventional level of morality is a good target for many of us. Only 20 to 25 percent of adults ever reach this stage, however.[15] At this stage people will obey laws if they serve the common good. No longer do people simply follow social convention or laws but actively seek to change them if those laws no longer promote the general welfare. At this stage, people may question the impact that materialistic values and behavior have on them, their loved ones, and the larger community. It is this level of moral reasoning that is required to opt out of the consumer culture, because people at this stage recognize the underlying moral purpose of laws and social customs.

If you reach stage 6 of Kohlberg's hierarchy of moral reasoning, you are on hallowed ground. Here moral thought is based on high-level thinking taking into account relevant universal truths. Laws are valid only if just.

A strong commitment to justice at this stage comes with an attendant obligation to disobey unjust laws or practices. Empathy is also important at this stage. Just decisions are arrived at only after considering others' points of view. Individuals act in a certain way because it is the right thing to do, not because it's what is expected of them, the easy way to go about things or sanctioned by law. Although Kohlberg theorized that stage 6 existed, it was difficult for him to pinpoint individuals who might consistently operate at this stage. Historical examples of those who might have reached stage 6 reasoning include Mahatma Gandhi, St. Francis of Assisi, Martin Luther King Jr., Mother Teresa of Calcutta, Albert Schweitzer, and Abraham Lincoln.

SHOPPING FOR MEANING

We can't all be Mother Teresa or Mahatma Gandhi, but each of us can live a life of meaning. Ultimately, materialism is an attempt (albeit misguided) to find meaning in one's life. This book, however, has shown that materialism robs us of the opportunity to have a meaningful life. It fills our days with shiny objects that give us fleeting pleasure and then pile up in our closets, garages, and storage units, only to be replaced with next year's model. Like a bulimic who purges after eating too much, we experience an endless binge-and-purge cycle of "things." Former Egyptian president Anwar El Sadat put it thusly, "Most people seek after what they do not possess and are thus enslaved by the very things they want to acquire."[16] The paradox lies in the fact that once we have captured and secured our prey, it loses much of its value. The ease with which we part with our belongings suggests that attachment is inversely related to quantity.[17] We care little for the things themselves; it's their *pursuit* we're after, because it keeps us distracted and helps us avoid the question we all dread: "Is this the life I want to live?"

We need to choose to live more simply, limiting our material consumption in order to free up our time and money to find happiness in nonmaterial pursuits. We must not allow our identity to be defined by the likes of Madison Avenue. This is not a rant against the free-enterprise system or an argument that we should all grow our hair long and live in communes.

It's about finding a balance between our desire for wealth and material possessions and our other life goals. Dare I say moderation in all things?

Some say we are suffering from the disease of "affluenza"—a bloated, sluggish, and unfulfilled feeling that comes from our dogged efforts to keep up with the Joneses.[18] It is a disease of epidemic proportions—in today's terms, a pandemic that has left in its wake a nation of overworked and stressed-out inhabitants in debt up to their ears and facing a pile of waste. Our souls do not hunger for wealth and possessions, even if that's what it sometimes feels like, but for meaning. As Kushner aptly puts it, "Our souls hunger for meaning, for the sense that we have figured out how to live so that our lives matter, so that the world will be at least a little bit different for our having passed through it."[19]

I am convinced that healthy relationships are the real "stuff" of human happiness. Essayist and poet Ralph Waldo Emerson hit the nail on the head when he said, "A Friend may well be reckoned the masterpiece of Nature."[20] We spend a lot of time worrying and rejoicing over friends won and lost, which is an indication of people's importance to us. Years later I can still remember kind words said to me by friends and family. At some level we all seem to understand that relationships are the key to our happiness.

Yet materialism separates us from others. It is a comparative process that sees others as competition instead of fellow human beings. And its pursuit comes at a great cost. How many dance recitals or ball games were missed or anniversaries forgotten because of work? How many divorces can be tied to out-of-control spending and fights over money? Why do we feel like we never measure up to others? Why do we no longer share a sense of community? Most of us would be hard-pressed to name our neighbors or—even less likely—state their birthdays.

Has the American Dream devolved into little more than a mad dash for material possessions? If it has, it is a sad thing, because I can think of nothing nobler than the opportunity to pursue a life of your choosing—the opportunity to live free from want while others do the same. Our founding fathers promised us the pursuit of happiness but left its definition up to us. It's time that we put away our selfish pursuits, get out from under our possessions, and rediscover the true meaning of happiness.

APPENDIX

"I LOVE MY STUFF": MEASURING MATERIALISM

Before we move on, each of us ready to take on and reform ourselves and the world, I thought you might like to see if you're a card-carrying member of the Shiny Objects Nation. Taking an eyes-wide-open self-assessment of where you stand on specific issues related to materialism is an important step in addressing the role materialism plays in your life.

What you see here is a commonly used scale to measure materialism. Read each statement and respond by circling the number that best represents the extent to which you agree or disagree with that statement. Be honest in answering each question. This is not graded homework, and no one else needs to know how you did. The results may be extremely beneficial and enlightening.

Materialism Scale

Circle the extent to which you agree or disagree with each question below.

	Strongly Agree	Agree	Neutral	Disgree	Strongly Disagree
1. I admire people who own expensive homes, cars, and clothes.	5	4	3	2	1

	Strongly Agree	Agree	Neutral	Disgree	Strongly Disagree
2. Some of the most important achievements in life include acquiring material possessions.	5	4	3	2	1
3. I don't place much emphasis on the amount of material objects people own as a sign of success.	5	4	3	2	1
4. The things I own say a lot about how well I'm doing in life.	5	4	3	2	1
5. I like to own things that impress people.	5	4	3	2	1
6. I try to keep my life simple, as far as possessions are concerned.	5	4	3	2	1
7. The things I own aren't all that important to me.	5	4	3	2	1
8. Buying things gives me a lot of pleasure.	5	4	3	2	1
9. I like a lot of luxury in my life.	5	4	3	2	1
10. I put less emphasis on material things than most people I know do.	5	4	3	2	1

	Strongly Agree	Agree	Neutral	Disgree	Strongly Disagree
11. I have all the things I really need to enjoy life.	5	4	3	2	1
12. My life would be better if I owned certain things I don't have.	5	4	3	2	1
13. I wouldn't be any happier if I owned nicer things.	5	4	3	2	1
14. I'd be happier if I could afford to buy more things.	5	4	3	2	1
15. It sometimes bothers me quite a bit that I can't afford to buy all the things I'd like.	5	4	3	2	1

SCORING INSTRUCTIONS

1. *Add up your scores for items 1, 2, 4, 5, 8, 9, 11, 12, 14, and 15. Your subtotal:*

   _____

2. *Reverse your score for items 3, 6, 7, 10, and 13. That is, if you scored a 5, give yourself a 1; if you scored a 4, give yourself a 2; etc.*

3. *Your subtotal:* _____

4. *Add your subtotals from steps 1 and 2. This total represents your materialism score and should range between 15 and 75.*

Write your materialism score here: _____

The following benchmarks are based upon a recent survey conducted with 403 adults from across the United States. Survey respondents ranged in age from eighteen to sixty-five and completed the materialism scale as part of a larger survey. Let's see how you stack up against this sample of typical consumers.

If you scored below 42, you are in the bottom 50 percent of the sample in terms of materialism (the average materialism score was 41.3). If you scored less than 35, you are in the bottom 25 percent of the sample (a sample chosen to mirror the U.S. population). On the flip side, if you scored in the range of 36 through 45, you are in the middle 50 percent of the sample. If you scored above 46, you are in the top 25 percent of U.S. adults regarding materialism. Is there a Betty Ford clinic for materialism? There should be.

Source: Marsha L. Richins and Scott Dawson, 1992, "A Consumer Values ori-entation for materialism and its measurement: Scale development and validation," Journal of Consumer Research 19 (December), 303-316.

"I CAN'T HELP MYSELF": MEASURING SELF-CONTROL

Presented here is a commonly used scale to measure self-control. Read each statement and respond by circling the number that best represents your typical state. Again, be as honest as you can be. For this assessment tool to be accurate—indeed, for this book to be effective—you need to take a comprehensive, frank look at your attitudes and behavior.

Self-Control Scale

Circle the number that best describes how much each of the following statements describes how you typically feel about yourself.

	Not at all				Very much
1. I am good at resisting temptation.	1	2	3	4	5
2. I have a hard time breaking bad habits.	1	2	3	4	5
3. I am lazy.	1	2	3	4	5
4. I say inappropriate things.	1	2	3	4	5

	Not at all				Very much
5. I do certain things that are bad for me, if they are fun.	1	2	3	4	5
6. I refuse things that are bad for me.	1	2	3	4	5
7. I wish I had more self-discipline.	1	2	3	4	5
8. People would say that I have iron self-discipline.	1	2	3	4	5
9. Pleasure and fun sometimes keep me from getting work done.	1	2	3	4	5
10. I have trouble concentrating.	1	2	3	4	5
11. I am able to work effectively toward long-term goals.	1	2	3	4	5
12. Sometimes I can't stop myself from doing something, even if I know it is wrong.	1	2	3	4	5
13. I often act without thinking through all the alternatives.	1	2	3	4	5

SCORING INSTRUCTIONS

1. *Add up your scores for items 1, 6, 8, and 11. Your subtotal:* _____

2. *Reverse your score for items 2–5, 7, 9, 10, 12, and 13. That is, if you scored a 5, give yourself a 1; if you scored a 4, give yourself a 2; etc.*

3. *Your subtotal:* _____

4. *Add your subtotals from steps 1 and 2. This total represents your self-control score and should range between 13 and 65.*

Write your self-control score here: _____

Based upon a recent sample of 403 adults from across the United States (ranging in age from eighteen to sixty-five) who completed this self-control scale as part of a larger survey, here are the benchmarks for seeing how you stack up.

If you scored below 44, you are in the bottom 50 percent of the sample in terms of self-control (the average self-control score was 44). If you scored less than 38, you are in the bottom 25 percent of the sample (a sample chosen to mirror the U.S. population). On a brighter note, if you scored in the range of 44–65, you are in the top 50 percent of the sample. Finally, if you scored above 50, you are in the top 25 percent of adults regarding self-control—and you should consider writing a book.

Source: June P. Tangney, Roy F. Baumeister, and Angie Luzio Boone, 2004, "High self-control predicts good adjustment, less pathology, better grades, and interpersonal success," Journal of Personality *72 (2), 271-324.*

NOTES

CHAPTER 1: SHINY OBJECTS

1. www.uncyclopedia.com/wiki/shiny-things, accessed July 17, 2009. I used the phonetic spelling for it from the words *shiny* and *object* from the American Heritage Dictionary.
2. Diane Brady and Christopher Palmeri, "The Pet Economy: Americans Spend an Astonishing $41 Billion a Year on Their Furry Friends," *Business Week*, August 6, 2007, 44; www.gushmagazine.com, accessed June 6, 2009; Kevin Helliker, "The Rise of Beds, and Fall of Dogs," *Wall Street Journal*, December 12, 2008, W8.
3. This phrase is similar to a title of a blog written by Baltimore Sun business reporter Eileen Ambrose in December 7, 2009.
4. George Ritzer, *Expressing America: A Critique of the Global Credit Card Society* (Thousand Oaks, CA: Pine Forge Press, 1995); Jathon Sapsford, "Paper Losses: As Cash Fades, America Becomes a Plastic Nation," *Wall Street Journal*, July 23, 2004, A1.
5. Russell W. Belk, 1988, "Third World Consumer Culture," in *Marketing and Development: Towards Broader Dimensions*, Erdogan Kumcu and A. Fuat Firat, eds. (Greenwich, CT: JAI Press), 103–127.
6. Robert D. McFadden and Angela MacroPoulos, "Wal-Mart Employee Trampled to Death," *New York Times*, May 29, 2008; "Wal-Mart Worker Trampled to Death Lacked Training, Attorney Says," December 1, 2008, www.foxnews.com/printer_friendly_story10,3566,459844,00.html.
7. Tim Kasser, *The High Price of Materialism* (Cambridge, MA: MIT Press, 2002).
8. Robert Triest, New England Economic Adventure, www.economic adventure.org, accessed June 8, 2009.

9. James Allan Davis and Tom W. Smith, General Social Surveys, 1972–2008, www.norc.org/gsst, accessed September 15, 2009.

10. Kasser, *The High Price of Materialism.*

11. National Payroll Week, "Getting Paid in America", Survey Results, 2008.

12. Kathleen D. Vohs, Nicole L. Mead, and Miranda R. Goode, "Merely Activating the Concept of Money Changes Personal and Interpersonal Behavior," *Current Directions in Psychological Science*, 17 (2008): 208–212.

13. Dan N. Stone, Ben Wier, and Stephanie M. Bryant, "Reducing Materialism Through Financial Literacy," *CPA Journal*, February 2008, 12–14.

14. Marvin E. Goldberg et al., "Understanding Materialism Among Youth," *Journal of Consumer Psychology* 13, no. 3 (2003): 278–288; "Parents and Media," Henry J. Kaiser Family Foundation, www.education.com/reference/article/ref_parents_media, accessed January 5, 2010.

15. J. H. Pryor et al., *The American Freshman: National Norms Fall 2008* (Los Angeles: Cooperative Institutional Research Program, Higher Education Research Institute, UCLA, 2009).

16. *The 2009 MetLife Study of the American Dream*, 4.

17. American Bankruptcy Institute, www.abiworld.org, accessed January 5, 2010.

18. www.cardweb.com, accessed with membership pass in January 2009.

19. Laura Thomas, "Consumer Culture Turns into Murals of Trash," *San Francisco Chronicle*, September 27, 2008.

20. Victor Lebow, "Price Competition in 1955," *Journal of Retailing* 31, no. 1: 7.

21. George W. Bush, *New York Times*, October 12, 2001, B4.

22. Jamie Arndt, Shelodon Solomon, Tim Kasser, and Kennon M. Sheldon, "The Urge to Splurge: A Terror Management Account of Materialism and Consumer Behavior," *Journal of Consumer Psychology*, 14, no. 3 (2004): 198–212.

23. "Brandcameo-films," www.brandchannel.com/brandcameo-films.asp, accessed June 19, 2009.

24. Maria Esposito, "Will Advertainment Be Curbed in the United States?" *The World*, July 11, 2008, 19.

25. W. Mischel, "From Good Intentions to Will Power," in *The Psychology of Action*, ed. P. Gollwitzer and J. Bargh, 197–218 (New York: Guilford, 1996).

26. *The 2009 MetLife Study of the American Dream*, 16.

27. Brian Orchard, "Daily Devotion: The Legacy of the Protestant Work Ethic," Fall 2004, www.vision.org/visionmedia/article.aspx?id=748, accessed December 12, 2009.

28. Thomas L. Friedman, "Why How Matters," *New York Times*, October 15, 2008, 35.

CHAPTER 2: CHASING THE AMERICAN DREAM

1. Brian Orchard, "Daily Devotion: The Legacy of the Protestant Work Ethic."
2. Roger B. Hill, "Protestantism and the Protestant Ethic," *History of Work Ethic*, 1996, www.coe.uga.edu/workethic/hpro.html, accessed September 14, 2009.
3. Steven Malanga, "Whatever Happened to the Work Ethic?" *City Journal*, Summer 2009, www.city-journal.org/printable.phy?id=5106, accessed September 10, 2009. The information about de Tocqueville and *Democracy in America* is also from this source.
4. Steven Malanga, "'Fatal Circle' of Materialism," *Waco Tribune Herald*, September 7, 2009, 6A; Max Weber, *The Protestant Work Ethic and the Spirit of Capitalism* (New York: Penguin Books, 2002 translation).
5. www.wikipedia.org/wiki/the_Protestant_Ethic_and_the_spirit_of_capitalism, accessed 12-18-2008.
6. Jim Cullen, *The American Dream: A Short History of an Idea That Shaped a Nation* (Oxford: Oxford University Press, 2003), 38.
7. Cullen, *American Dream*, 45.
8. Economic Mobility Project: An Initiative of the Pew Charitable Trusts, "Findings from a National Survey and Focus Groups on Economic Mobility," March 12, 2009, www.pewtrusts.org/uploadedFiles/wwwpew trustsorg/Reports/Economic_Mobility/EMP%202009%20Survey%20 on%20Economic%20Mobility%20FOR%20PRINT%203.12.09.pdf.
9. Malanga, "Whatever Happened to the Work Ethic?"
10. "Gold Fever: Discovery, the 48ers and the Rush of '49," www.learncalifornia .org/doc.asp?id=1928, accessed September 22, 2009.
11. "Gold Fever: Discovery, the 48ers, and the Rush of '49."
12. "Fever," www.isu.edu/~trinnich/fever.html, accessed on September 22, 2009.
13. "The California Gold Rush, 1849," *Eyewitness to History*, 2003, www.eyewitness tohistory.com/californiagoldrush.htm, accessed September 21, 2009.
14. H. W. Brands, *The Age of Gold* (New York: Doubleday, 2002), 442.
15. T. J. Jackson Lears, "From Salvation to Self-Realization," in *The Culture of Consumption*, ed. R. W. Fox and T. J. Jackson Lears (New York: Pantheon Books, 1983), 3.
16. Gary Cross, *An All-Consuming Century* (New York: Columbia University Press, 2000), 21.
17. Cross, *All-Consuming Century*, 28.

18. Richard S. Tedlow, *New and Improved: The Story of Mass Marketing in America* (New York: Basic, 1990).

19. Cross, *All-Consuming Century*, 21, 31.

20. Cross, *All-Consuming Century*, 29.

21. William Leach, *Land of Desire* (New York: Pantheon Books, 1993), 3.

22. Leach, *Land of Desire*, 3.

23. Yiannis Gabriel and Tim Lang, "New Faces and New Masks of Today's Consumer," *Journal of Consumer Culture* 8, no. 3 (2008): 323.

24. Gabriel and Lang, "New Faces and New Masks," 323.

25. Cross, *All-Consuming Century*, 26.

26. Leach, *Land of Desire*, 266.

27. In Gabriel and Lang, "New Faces and New Masks," 326.

28. Cross, *All-Consuming Century*, 62.

29. The statistics in this paragraph and the next are from Cross, *All-Consuming Century*.

30. Leach, *Land of Desire*, 267.

31. James Truslow Adams, *The Epic of America* (Boston: Little, Brown, 1931).

32. Cullen, *American Dream*, 4.

33. Adams, *Epic of America*, viii.

34. Cullen, *American Dream*, 7.

35. Adams, *Epic of America*, 404.

36. Adams, *Epic of America*, 404.

37. Adams, *Epic of America*, 406.

38. Adams, *Epic of America*, 411.

39. David Kamp, "Rethinking the American Dream," *Vanity Fair*, April 2009, www.vanity-fair.com/culture/features/2009/09/American-Dream200904, accessed August 19, 2009.

40. Kamp, "Rethinking the American Dream."

41. Kamp, "Rethinking the American Dream."

42. Franklin D. Roosevelt, "The Four Freedoms," message delivered to Congress January 6, 1941, www.americanrhetoric.com/speeches/fdrthefourfreedoms.html, accessed September 28, 2009.

43. www.todaysamericandream.com, accessed May 20, 2010.

44. Roosevelt, "Four Freedoms."

45. "Norman Rockwell's Four Freedoms in the *Saturday Evening Post*, February 20 to March 13," *Best Norman Rockwell Art*, September 28, 2009, www.best-norman-rockwell-art.com/fourfreedoms.html, accessed September 28, 2009.

46. Kamp, "Rethinking the American Dream."

47. "Norman Rockwell's Four Freedoms."

CHAPTER 3: THE AMERICAN DREAM ON STEROIDS

1. The two Nixon quotes come from R. W. Fox and T. J. Jackson Lears, eds., *The Culture of Consumption* (New York: Pantheon Books, 1983), ix–x.

2. Cross, *All-Consuming Century*, 84.

3. Cross, *All-Consuming Century*, 84.

4. Cross, *All-Consuming Century*, 87.

5. Kamp, "Rethinking the American Dream."

6. Cross, *All-Consuming Century*, 95.

7. Jan Logemann, "Consumerism and Leisure," *Journal of Social History*, Spring 2008, 543.

8. In David Steigerwald, "Did the Protestant Ethic Disappear? The Virtue of Thrift on the Cusp of Postwar Affluence," *Business History Conference* (Oxford: Oxford University Press, 2008), 806.

9. Cross, *All-Consuming Century*, 92.

10. Cross, *All-Consuming Century*, 100.

11. Cross, *All-Consuming Century*, 100.

12. Cross, *All-Consuming Century*, 100.

13. Cross, *All-Consuming Century*, 103.

14. Kamp, "Rethinking the American Dream."

15. Kamp, "Rethinking the American Dream."

16. Steigerwald, "Did the Protestant Ethic Disappear?" (page number TK)

17. Steigerwald, "Did the Protestant Ethic Disappear?" 806.

18. Logemann, "Consumerism and Leisure," 525.

19. Ernest Dichter, *Handbook of Consumer Motivations* (New York: McGraw-Hill, 1964), 161.

20. Steigerwald, "Did the Protestant Ethic Disappear?" 795.

21. Steigerwald, "Did the Protestant Ethic Disappear?" 793.

22. Steigerwald, "Did the Protestant Ethic Disappear?" 797.

23. www.spiritus-temporis.com/counterculture/19602-counterculture.html, accessed September 29, 2009; Cross, *All-Consuming Century*, 163.

24. Cross, *All-Consuming Century*, 152.

25. Herbert Marcuse, *One-Dimensional Society* (Boston: Beacon Press, 1964).

26. Juliet Schor, *The Overspent American* (New York: Harper Perennial, 1998).

27. Charles Reich, *The Greening of America* (New York: Random House, 1970), 231; Jerry Rubin, *Do It: Scenarios of a Revolution* (New York: Ballantine 1970), 87.

28. Jerry Rubin, *Growing (Up) at Thirty-Seven* (New York: Evans, 1976), 20–21.

29. Godfrey Hodgson, *America in Our Time* (Doubleday, 1976), 365.

30. Cross, *All-Consuming Century*, 178.

31. Cross, *All-Consuming Century*, 179-183.

32. Cross, *All-Consuming Century*, 170-171.

33. Cross, *All-Consuming Century*, 170.

34. Logemann, "Consumerism and Leisure," 528.

35. Logemann, "Consumerism and Leisure," 535.

36. Louis Hyman, "Debtor Nation: How Consumer Credit Built Postwar America," *Business History Conference* (Oxford Press, 2008), 615.

37. Mark J. Perry, "The Worst Economy Since . . . the 1980s?" *Carpe Diem*, September 25, 2009, www.mjperry.blogspot.com, accessed October 15, 2009. The statistics in the paragraph that follows are also from Perry's blog.

38. "People and Events: Carter's 'Crisis of Confidence' Speech," www.pbs.org/wgbh/amex/carter/peopleevents/e_malaise.html, accessed October 8, 2009.

39. "People and Events: Carter's 'Crisis of Confidence' Speech."

40. James Carter, "The 'Malaise Speech' 1979," www.volstate.edu/geades/finaldocs/1970s+beyond/malaise.htm, accessed October 8, 2009.

41. Carter, "'Malaise Speech' 1979."

42. Carter, "'Malaise Speech' 1979."

43. Cross, *All-Consuming Century*, 162.

44. "A Time for Choosing," www.reagan2020.us/speeches/a_time_for_choosing.asp, accessed October 11, 2009.

45. John de Graaf, David Wann, and Thomas H. Naylor, *Affluenza* (San Francisco: Berrett-Koehler, 2005), 154.

46. De Graaf, Wann, and Naylor, *Affluenza*, 153.

47. Kamp, "Rethinking the American Dream."

48. Cullen, *American Dream*, 170.

49. *A Charlie Brown Christmas*, CBS TV, 1965.

50. www.wikipedia.org/wiki/mark_cuban, accessed October 13, 2009.

51. Eric Niiler, "Netscape's IPO Anniversary and the Internet Boom," National Public Radio, August 9, 2005, www.npr.org/templates/story/story.php?storyId=4792365, accessed October 13, 2009.

52. Index chart, www.dynamic.NASDAQ.com/dynamic/indexchart.asp?, accessed October 12, 2009.

53. "Top 10 dot-com flops," cnet.com, www.cnet.com/4520-11136_1-6278387-1.html, accessed October 17, 2009.

54. Jonathan R. Laing, "The Bubble's New Home," *Barron's*, June 20, 2005, www.online.barrons.com/article151311905372884363176.html.

55. Cullen, *American Dream*, 160.

56. Matthew Warshauer, "Who Wants to Be a Millionaire? Changing

Conceptions of the American Dream," February 13, 2003, www.americansc
.org.uk/online/american_dream.htm, accessed December 5, 2008.

57. Cullen, *American Dream*, 178.

58. Steven Malanga, "Obsessive Housing Disorder: Nearly a Century of Washington's Efforts to Promote Homeownership Has Produced One Calamity After Another. Time to Stop," *City Journal*, Spring 2009, www.city-journal.org/printable.php?id=4376, accessed September 14, 2009.

59. Thomas J. Sugrue, "The New American Dream: Renting," *Wall Street Journal*, August 14, 2009, www.online.wsj.com, accessed August 19, 2009.

60. Malanga, "Obsessive Housing Disorder."

61. Cullen, *American Dream*, 138.

62. Cullen, *American Dream*, 141.

63. Cullen, *American Dream*, 148.

64. Malanga, "Obsessive Housing Disorder."

65. Malanga, "Obsessive Housing Disorder."

66. Malanga, "Obsessive Housing Disorder."

67. Malanga, "Obsessive Housing Disorder."

68. Malanga, "Obsessive Housing Disorder."

69. Malanga, "Obsessive Housing Disorder."

70. Louis Hyman, as quoted in Malanga, "Obsessive Housing Disorder."

71. Malanga, "Obsessive Housing Disorder."

72. Malanga, "Obsessive Housing Disorder"; Sugrue, "New American Dream."

73. "The National Homeownership Strategy: Partners in the American Dream," U.S. Department of Housing and Urban Development, Washington, DC, 1995, www.udm.dalnet.lib.mi.us/ipac20/ipacJSP?session =S20/376151890.1310617+profile=dm, accessed on January 6, 2010; Michael S. Carliner, "Development of Federal Homeownership 'Policy,'" *Housing Policy Debate* 9, no. 2 (1998): 299–321.

74. Sugrue, "New American Dream."

75. Sugrue, "New American Dream."

CHAPTER 4: THE CAT'S OUT OF THE (SHOPPING) BAG

1. Ed Diener, Jeff Horowitz, and Robert Emmons, "Happiness of the Very Wealthy," *Social Indicators Research* 16, no. 3 (1985): 263–74.

2. Ed Diener and Martin E. P. Seligman, "Beyond Money: Toward an Economy of Well-Being," *Psychological Science in the Public Interest* 5 (2004): 1–31.

3. www.quotegarden.com, accessed December 15, 2009.

4. Bob Cumming, "Understanding Happiness," www.abc.net.au/news/stories/2008/05/29/2258813.htm, accessed September 19, 2009.

5. Ed Diener, Eunbrook M. Suh, Richard E. Lucas, and Heidi L. Smith, "Subjective Well-Being: Three Decades of Progress," *Psychological Bulletin* 125, no. 2 (1999): 276–302; Paul Dolan, Tessa Peasgood, and Mathew White, "Do We Really Know What Makes Us Happy? A Review of the Economic Literature on the Factors Associated with Subjective Well-Being," *Journal of Economic Psychology* 29 (2008): 94–122; Ed Diener and Robert Biswas-Diener, "Will Money Increase Subjective Well-Being?" *Social Indicators Research* 57 (2002): 119–69.

6. Thomas L. Friedman, "2008: When Mother Nature and the Market Both said: 'No More'," *New York Times*, March 10, 2009.

7. Ruth Davidhizar and April Hart, "Are You Born a Happy Person or Do You Have to Make It Happen?" *The Health Care Manager* 25, no. 1 (2006): 64–69.

8. Martin E. P. Seligman, *Authentic Happiness* (New York: Free Press, 2002), 45.

9. Sonja Lyubomirsky and H. S. Lepper, "A Measure of Subjective Happiness: Preliminary Reliability and Construct Validation," *Social Indicators Research* 46 (1999): 137–55.

10. Arthur C. Brooks, "Can Money Buy Happiness?" *The American*, May/June 2008, www.american.com, accessed October 17, 2008.

11. Richard A. Easterlin, "Feeding the Illusion of Growth and Happiness: A Reply to Hagerty and Veenhoven," *Social Indicators Research* 74 (2005): 429–43; Ed Diener and S. Oishi, "Money and Happiness: Income and Subjective Well-Being Across Nations," in *Subjective Well-Being Across Cultures* (Cambridge, MA: MIT Press, 2000); Diener and Seligman, "Beyond Money."

12. Diener and Biswas-Diener, "Will Money Increase Subjective Well-Being?"

13. Richard A. Easterlin, "Will Raising the Income of All Increase the Happiness of All?" *Journal of Economic Behavior and Organization* 27, no. 1 (1995): 35–47.

14. E-mail correspondence with Ed Diener, October 17, 2008, unpublished research.

15. Daniel Kahneman and Alan Krueger, "Developments in the Measurement of Subjective Well-Being," *Journal of Economic Perspectives* 20, no. 1 (2006): 3–24; Shankar Vedantam, "Financial Hardship and the Happiness Paradox," *Washington Post*, June 23, 2008, www.washingtonpost.com, accessed October 17, 2008.

16. Trevor Corson, "What Finland Can Teach America About True Luxury," *Christian Science Monitor*, May 1, 2009, www.csmonitor.com/2009/0501/p09502-coop.html, accessed May 5, 2009.

17. Russell W. Belk, "Materialism: Trait Aspects of Living in a Material World," *Journal of Consumer Research* 12 (1985): 265–80.

18. Kasser, *High Price of Materialism*.

19. James A. Roberts and Aimee Clement, "Materialism and Satisfaction with Over-All Quality of Life and Eight Life Domains," *Social Indicators Research* 82 (2007): 79–92.

20. Emily Solberg, Ed Diener, and Michael Robinson, "Why Are Materialists Less Satisfied?" in *Psychology and Consumer Culture: The Struggle for a Good Life in a Materialistic World*, ed. Tim Kasser and Allen D. Kanner (Washington, DC: American Psychological Association, 2004), 29–48.

21. Kasser, *High Price of Materialism*.

22. Gregg Easterbrook, "The Real Truth About Money," *Time*, January 17, 2005, A32.

23. Centers for Disease Control, "Behavioral Risk Factor Surveillance System," www.cdc.gov/mentalhealth/prevalence_data.html, accessed April 29, 2009.

24. Jean M. Twenge et al., "Birth cohort increases in psychopathology among young Americans, 1938-2007; a cross-temporal meta-analysis of the MMPI," *Clinical Psychology Review* 30 (2010): 145-154.

25. Oliver James, "Selfish Capitalism and Mental Illness," *Psychologist* 20, no. 7 (2007): 426–28.

26. James, "Selfish Capitalism and Mental Illness," 426–28.

27. Kasser, *High Price of Materialism*.

28. James, "Selfish Capitalism and Mental Illness," 426–28.

CHAPTER 5: THE TREADMILL OF CONSUMPTION

1. John E. Lycett and Robin I. M. Dunbar, "Mobile Phones as Lekking Devices Among Human Males," *Human Nature* 11, no. 1 (2000): 93–104.

2. Philip Brickman et al., "Lottery Winners and Accident Victims: Is Happiness Relative?" *Journal of Personality and Social Psychology* 36, no. 8 (1978): 917–27.

3. Darren Murph, "Motorola Intros Avant-Garde $2,000 Aura, Markets It Like a Rolex," October 21, 2008, www.endgadget.com, accessed October 21, 2008.

4. Jacqueline Eastman et al., "The Relationship Between Status Consumption and Materialism: A Cross Cultural Comparison of Chinese, Mexican, and

American Students," *Journal of Marketing Theory and Practice*, Winter 1997, 52–66, 58.

5. Ted Turner quote, www.quotegarden.com, accessed November 15, 2009.

6. Eastman et al., "Relationship Between Status Consumption and Materialism," 52–66, 58.

7. Ronald Faber and Thomas O'Guinn, "A Clinical Screener for Compulsive Buying," *Journal of Consumer Research* (1992): 459–469.

8. James A. Roberts and Carlos Martinez, "The Emerging Consumer Culture in Mexico: An Explanatory Investigation of Compulsive Buying in Mexican Young Adults," *Journal of International Consumer Marketing* 10 (1997): 7–31.

9. James A. Roberts, "Compulsive Buying Among College Students: An Investigation of Its Antecedents, Consequences, and Implications for Public Policy," *Journal of Consumer Affairs* 32 (1998): 295–319.

10. Roberts, "Compulsive Buying Among College Students."

11. B. Mittal and others, *Consumer Behavior: How Humans Think, Feel, and Act in the Marketplace* (Cincinnati, OH: Open Mentis Publishing, 2008).

12. James A. Roberts and Jeff Tanner, "Compulsive Buying and Risky Behavior Among Adolescents," *Psychological Reports* 86 (2000): 763–70.

13. James A. Roberts and Eli Jones, "Money Attitudes, Credit Card Use, and Compulsive Buying Among American College Students," *Journal of Consumer Affairs* 35, no. 2 (2001): 213–40.

14. James A. Roberts and Stephen Pirog, "Personal Goals and Their Role in Consumer Behavior: The Case of Compulsive Buying," *Journal of Marketing Theory and Practice* 12, no. 3 (2004): 61–73.

15. Lorrin M. Koran et al., "Estimated Prevalence of Compulsive Buying in the United States," *The American Journal of Psychiatry*, October 2006, 1806–12.

16. Chris Manolis, James A. Roberts and Vishal Kashyap, "A Critique and Comparison of Two Scales from Fifteen Years of Studying Compulsive Buying," Psychological Reports 102 (2008): 153-165.

17. Lukas R. Dean, Jason S. Carroll, and Chongming Yang, "Materialism, Perceived Financial Problems, and Marital Satisfaction," *Families and Consumer Sciences Research Journal* 35 (2007): 260–81.

18. "Grieving 9/11 Widow Spends Almost $5 Million," *ABC News*, June 13, 2005, www.abcnews.go.com/erin/?id=843920, accessed May 2, 2008.

19. Ronald Faber, "Self-Control and Compulsive Buying," in *Psychology and Consumer Culture*, ed. Tim Kasser and Allen D. Kanner (Washington, DC: American Psychological Association, 2004), 169–87.

20. Ronald Faber and Thomas O'Guinn, "A Clinical Screener for Compulsive Buying."

CHAPTER 6: THE CASHLESS SOCIETY

1. This text, slightly edited, is taken from www.debtorsanonymous.org/help/signs.htm, accessed October 24, 2009.

2. www.bea.gov/briefrm/saving.html, accessed January 5, 2010.

3. Laura Conaway, "Household Debt vs. GDP," *Planet Money*, May 13, 2009, www.npr.org/blogs/money/2009/02/household_debt_vs_gdp.html, accessed May 13, 2009.

4. www.creditcards.com/credit-card-news/credit-card-industry-facts-personal-debt-statistics-1276.php, accessed June 20, 2009.

5. Bruce Krasting, "US Mortgage Market 2000–2008—Follow the Money," April 23, 2009, www.brucekrasting.blogspot.com/2009/04/us-mortgage-market-2000-2008-follow.html, accessed December 15, 2009.

6. Gretchen Morgenson, "The Debt Trap," *New York Times*, July 20, 2008, www.nytimes.com/2008/07/20/business/20debt.html, accessed May 13, 2009.

7. www.creditcards.com/credit-card-news/credit-card-industry-facts-personal-debt-statistics-1276.php, accessed June 21, 2009.

8. www.creditcards.com/credit-card-news/credit-card-industry-facts-personal-debt-statistics-1276.php.

9. www.creditcards.com/credit-card-news/credit-card-industry-facts-personal-debt-statistics-1276.php.

10. www.creditcards.com/credit-card-news/credit-card-industry-facts-personal-debt-statistics-1276.php.

11. Scott I. Rick, Cynthia E. Cryder, and George Lowenstein, "Tightwads and Spendthrifts," *Journal of Consumer Research* 34 (2008): 767–82.

12. John J. Watson, "The Relationship of Materialism to Spending Tendencies, Saving, and Debt," *Journal of Economic Psychology* 24, no. 6 (2003): 723–39.

13. Julie Fitzmaurice, "Splurge Purchases and Materialism," *Journal of Consumer Marketing* 25 (2008): 332–38.

14. M. Joseph Sirgy, "Materialism and Quality of Life," *Social Indicators Research* 43 (1998): 227–60.

15. Jennifer Agiesta and Jon Cohen, "Financial Anxiety Rising, Poll Finds," *Washington Post*, October 13, 2008, A13.

16. Jing Jian Xiao, Chuanyi Tang, and Soyeon Shim, "Acting for Happiness:

Financial Behavior and Life Satisfaction of College Students," *Social Indicators Research* 92 (2009): 53–68.

17. Sapsford, "Paper Losses."

18. www.truecostofcredit.com1549035, accessed June 2009.

19. Richard A. Feinberg, "Credit Cards as Spending Facilitating Stimuli: A Conditioning Interpretation," *Journal of Consumer Research* 13 (1986): 348–56.

20. Dilip Soman, "Effects of Payment Mechanism on Spending Behavior: The Role of Rehearsal and Immediacy of Payment," *Journal of Consumer Research* 27 (2001): 460–74.

21. Sallie Mae (SLM Corporation), "How Undergraduate Students Use Credit Cards," 2009, 3.

22. "Student Credit Card Debt," *Commercial Law Bulletin* 13 (1998): 6–7.

CHAPTER 7: MONEY'S HIDDEN COSTS

1. Pryor et al., *American Freshman.*

2. David G. Myers, *Psychology* (New York: Worth Publishers, 2004), 587–90; Kasser, *High Price of Materialism.*

3. Kasser, *High Price of Materialism.*

4. Peter Gray, *Psychology*, 5th ed. (New York: Worth Publishers, 2007), 565–66.

5. Tim Kasser and Richard Ryan, "Further Examining the American Dream: Differential Correlates of Intrinsic and Extrinsic Goals," *Journal of Personality and Social Psychology*, 1996, 280–87; Tim Kasser and Richard Ryan, "A Dark Side of the American Dream: Correlates of Financial Success as a Central Life Aspiration," *Journal of Personality and Social Psychology*, 1993, 410–22.

6. Pew Research Center, "A Portrait of 'Generation Next,'" January 9, 2007, www.peoplepress.org/report/300/a-portrait-of-generation-next, accessed December 7, 2009.

7. Kasser and Ryan, "Further Examining the American Dream"; Kasser and Ryan, "A Dark Side of the American Dream."

8. Megan Basham, "Bringing Up Princess: Turning Girls into Narcissists," *Wall Street Journal*, June 12, 2009, W13.

9. Patricia Cohen and Jacob Cohen, *Life Values and Adolescent Mental Health* (Hillsdale, NJ: Erlbaum, 1996), 139.

10. Carol Nickerson, Norbert Schwarz, Ed Diener, and Daniel Kahneman, "Zeroing In on the Dark Side of the American Dream," *Psychological Science* 19, no. 6 (2003): 531–36.

11. Roberts and Pirog, "Personal Goals and Their Role in Consumer Behavior," 61–73.
12. Ruth Engs, Indiana University, "How Can I Manage Compulsive Shopping and Spending Addiction," www.indiana.edu/~engs/hints/shop.html, accessed January 17, 2010.

CHAPTER 8: COLLATERAL DAMAGE: RELATIONSHIPS

1. Philip Brickman et al., "Lottery Winners and Accident Victims: Is Happiness Relative?"
2. Diener and Biswas-Diener, "Will Money Increase Subjective Well-being?" 119–69.
3. Kasser, *High Price of Materialism*, 161.
4. Simone Pettigrew, "Consumption and the Ideal Life," January 31, 2006, www.jrconsumers.com, accessed January 31, 2006.
5. Pettigrew, "Consumption and the Ideal Life."
6. David M. Buss, "The Evolution of Happiness," *American Psychologist* 55, no. 1 (2000): 15–23, 17.
7. Ed Diener and Martin E. P. Seligman, "Very Happy People," *Psychological Science* 13 (2002): 80–83.
8. Shalom H. Schwartz, "Are There Universal Aspects in the Structure and Content of Human Values?" *Journal of Social Issues* 50 (1994): 19–45; James E. Burroughs and Aric Rindfleisch, "Materialism and Well-Being: A Conflicting Values Perspective," *Journal of Consumer Research* 29, no. 3 (2002): 348–370; Kasser, *High Price of Materialism*.
9. Thorstein Veblen, *The Theory of the Leisure Class* (New York: Random House, 1934).
10. Yiannis Gabriel and Tim Lang, "New Faces and New Masks of Today's Consumer," *Journal of Consumer Culture* 8, no. 3 (2008): 322.
11. Zygmunt Bauman, "Collateral Casualties of Consumerism," *Journal of Consumer Culture* 7, no. 1 (2007): 25–56.
12. Barry Schwartz, *The Costs of Living: How Market Freedom Erodes the Best Things in Life* (New York: Norton, 1994), 201.
13. Kasser, *High Price of Materialism*, 70.
14. Belk, "Materialism," 765–80.
15. Marsha Richins and Scott Dawson, "A Consumer Values Orientation for Materialism and Its Measurement: Scale Development and Validation," *Journal of Consumer Research* 19 (1992): 303–16.

16. Elizabeth W. Dunn, Lara B. Akrin, and Michael I. Norton, "Spending Money on Others Promotes Happiness," *Science*, March 21, 2008, 1687–88.

17. Kathleen D. Vohs, Nicole L. Mead, and Miranda R. Goode, "Merely Activating the Concept of Money Changes Personal and Interpersonal Behavior," *Current Directions in Psychological Science* 17 (2008): 208–12.

18. Dean, Carroll, and Yang, "Materialism, Perceived Financial Problems, and Marital Satisfaction."

19. Charlotte Gill, "£44,500 Shopaholic Ruins Her Parents," *Daily Mail*, January 4, 2005, 3; Natasha Courtenay-Smith, "Destroyed by a shopaholic daughter," *Daily Mail*, January 7, 2005, 60.

20. Eirini Flouri, "Exploring the Relationship Between Mothers' and Fathers' Parenting Practices and Children's Materialist Values," *Journal of Economic Psychology* 25 (2004): 743–52.

21. Kasser, *High Price of Materialism*, 32.

22. Paul R. Amato and Stacy J. Rogers, "A Longitudinal Study of Marital Problems and Subsequent Divorce," *Journal of Marriage and the Family* 59, no. 3 (1997): 612–24.

23. Barbara Brandt, "An Issue for Everybody," in *Take Back Your Time* (San Francisco: Berrett-Koehler, 2003), 12–19.

24. Norman Herr, "Television and Health," www.csun.edu/science/health/docs/tv&health.html, accessed January 6, 2010; Annie Leonard, "The Story of Stuff," www.storyofstuff.com, accessed February 12, 2010.

25. Arlie Hochschild, *The Commercialization of Intimate Life* (Berkeley: University of California Press, 2003), 203.

26. William Doherty and Barbara Carlson, "Overscheduled Kids, Underconnected Families," in *Take Back Your Time* (San Francisco: Berrett-Koehler, 2003), 38–45.

CHAPTER 9: WHY ARE WE SO MATERIALISTIC?

1. Kennon M. Sheldon and Tim Kasser, "Psychological Threat and Extrinsic Goal Striving," *Motivation and Emotion* 32 (2008): 37–45.

2. Sami Beg, "Mass Media Exposure," *ABC News*, December 15, 2006, www.abcnews.go.com/print?id=2727587, accessed May 16, 2008.

3. Herr, "Television and Health."

4. Moniek Buijzen and Patti M. Valkenburg, "The Unintended Effects of Television Advertising," *Communication Research* 30, no. 5 (2003): 483–503.

5. L. J. Shrum, James E. Burroughs, and Aric Rindfleisch, "Television's Cultivation of Material Values," *Journal of Consumer Research* 32, no. 3 (2005): 473–79.

6. James A. Roberts, Chris Manolis, and Jeff Tanner, "Interpersonal Influence and Adolescent Materialism and Compulsive Buying," *Social Influence* 3, no. 2 (2008): 114–31.

7. Marti Attoun, "Cool, Crazy, and Uncommon Collections," *American Profile* (newspaper magazine insert), January 4–10, 2009, 4–7, www .americanprofile.com/article/31094.html.

8. Tim Kasser and Kennon M. Sheldon, "Of Wealth and Death: Materialism, Mortality Salience, and Consumption Behavior," *Psychological Science*, July 4, 2000, 348–51.

9. Tim Kasser and Virginia Grow Kasser, "The Dreams of People High and Low in Materialism," *Journal of Economic Psychology* 22 (2001): 693–719.

10. Kasser, *High Price of Materialism*, 39.

11. Elizabeth C. Hirschman and Barbara B. Stern, "Do Consumers' Genes Influence Their Behavior?" *Advances in Consumer Research* 28 (2001): 403–10.

12. David M. Buss, "Sex Differences in Human Mate Preferences: Evolutionary Hypothesis Tested in 37 Cultures," *Behavioral and Brain Sciences* 12 (1989): 1–49.

13. Bob Cummins, "Understanding Happiness," *ABC News*, May 29, 2008, www.abc.net.au/news/stories/2008/05129/2258813.html, accessed June 24, 2008.

14. John C. Mowen, *The 3M Model of Motivation and Personality: Theory and Empirical Application to Consumer Behavior* (New York: Kluwer Academic Publishers, 2000).

15. Sharon Begley, "Parents Can Counteract 'Environments' Created by Children's Genes," *Wall Street Journal*, February 24, 2006, B1.

16. Marvin Zuckerman, "The Shaping of Personality: Genes, Environments, and Chance Encounters," *Journal of Personality Assessment* 82 (2004): 11–22.

17. Martin E. P. Seligman, *What You Can Change and What You Can't: The Complete Guide to Successful Self-Improvement* (New York: Knopf, 1994), 231.

18. Mowen, *3M Model of Motivation and Personality*.

19. Stephen F. Pirog and James A. Roberts, "Personality and Credit Card Misuse Among College Students," *Journal of Marketing Theory and Practice* 15, no. 1 (2007): 65–77.

20. Hirschman and Stern, "Do Consumers' Genes Influence Their Behavior?" 403–10.

21. Marvin Zuckerman, *Psychobiology of Personality* (Cambridge, UK:

Cambridge University Press, 2005); David J. Burns and Steven D. Burns, "Compulsive Behavior in the Marketplace and in the Workplace: An Examination of Causal Factors," in *Advances in Genetics Research*, ed. Kevin V. Urbano (London: Nova Science, 2010).

22.   Marvin Zuckerman, *Sensation Seeking: Beyond Optimal Level of Arousal* (Hillsdale, NJ: Lawrence Erlbaum Associates, 1980).

23.   Zuckerman, "The Shaping of Personality," 11–22.

CHAPTER 10: HEAVEN HELP US: THE PROSPERITY GOSPEL

1.   Jonathan L. Walton, "RD Quiz: Prosperity Gospel Self-Examination," April 15, 2009, www.religiondispatches.org/archive/religionandtheology/1319/rdquiz:prosperity_gospel_selfexamination, accessed June 17, 2009.

2.   "The Prosperity Gospel," *The First Post* (UK), May 27, 2009, www.thefirstpost.co.uk/45490,news-comment,news-politics,the-prosperity-gospel, accessed May 27, 2009.

3.   Paul Gifford, "The Prosperity Gospel in Africa," *Christian Century*, July 10, 2007, 20–24.

4.   Ginger Stickney, "Godly Riches: The 19th Century Roots of Modern Prosperity Gospel," in *Religion and Class in America: Culture, History, and Politics* (Brill, 2009), 161–62; Milmon F. Harrison, *Righteous Riches: The Word of Faith Movement in Contemporary African-American Religion* (New York: Oxford University Press, 2005).

5.   "Biography of Famous Preacher and Evangelist Reverend Ike: Part 1," www.trivia-library-com/a/biography-of-famous-preacher-and-evangelist-ike-par-1.htm, accessed 7/24/2009.

6.   Kate Taylor, "Rev. Ike Opens His Palace," *New York Sun*, March 6, 2007, www.NYSun.com/arts/rev-ike-opens-his-palace/49840/, accessed May 27, 2009.

7.   Richard N. Ostling, "Jim Bakker's Crumbling World," *Time*, December 19, 1988, www.time.com/time/magazine/article/0,9171,956551,00.html, accessed June 2, 2009.

8.   Jim Bakker, *I Was Wrong* (Nashville: Thomas Nelson, 1996), 535.

9.   Armen Keteyian, "On the Road to Prosperity," *CBS News*, January 29, 2008, www.cbsnews.com/blogs/2008/01/29/pr.marysource/entry37677445html, accessed November 13, 2008.

10.   "The 10 Most Fascinating People of 2006," *ABC News*, December 12, 2006, www.abcnews.go.com/2020/story?id=2716887&page=2; "The 50 Most Influential Christians in America," The Church Report, January 2007; "New Osteen Book at Three Million," *Publisher's Weekly*, April 13, 2007; Dr.

Terry Wakins, "Joel Osteen: True or False?", www.av1011.org/osteen.html, accessed November, 13, 2008.

11. "Joel Osteen Answers His Critics," *CBS News*, October 14, 2007, wwwcbsnews.com/stories/2007/10111/60minutes/man3358652.5html, accessed May 27, 2009.

12. Allan Greenblatt and Tracie Powell, "Rise of Megachurches," *CQ Researcher*, September 21, 2007, 769–91; "T.D. Jakes," www.en.wikipedia.org/wiki/T.D._Jakes, accessed June 25, 2008.

13. Greenblatt and Powell, "Rise of Megachurches," 782–83; "Creflo Dollar," www.en.wikipedia.org/wiki/creflo_dollar, accessed November 13, 2008.

14. Robert M. Franklin, "The Gospel of Bling," *Sojourners*, January 2007, 19.

15. David Van Biema, "Maybe We Should Blame God for the Subprime Mess," *Time*, October 3, 2008, www.time.com/time/business/article/0,8599,1847053,00.html, accessed May 27, 2009.

16. Stickney, "Godly Riches," 159.

17. George M. Thomas, *Revivalism and Cultural Change* (Chicago: University of North Carolina Press, 1989) in Stickney, "Godly Riches," 160.

18. Stickney, "Godly Riches," 163.

19. Stickney, "Godly Riches," 166.

20. Stickney, "Godly Riches," 166.

21. Stickney, "Godly Riches," 174.

22. Greenblatt and Powell, "Rise of Megachurches," 784.

23. Greenblatt and Powell, "Rise of Megachurches," 772.

24. David Van Biema and Jeff Chu, "Does God Want You to Be Rich?" *Time*, September 10, 2006, www.time.com/time/magazine/article/0,9171,1533448,00.html, accessed May 27, 2009.

25. Greenblatt and Powell, "Rise of Megachurches," 772.

26. Rodney Stark, "What Americans Really Believe," *Baylor University's Institute for Studies of Religion*, 2008, 46.

27. Van Biema and Chu, "Does God Want You to Be Rich?"

28. McFarland as quoted in Greenblatt and Powell, "Rise of Megachurches," 774.

29. Greenblatt and Powell, "Rise of Megachurches," 774.

30. Greenblatt and Powell, "Rise of Megachurches," 771.

31. Greenblatt and Powell, "Rise of Megachurches," 771.

32. Ross as quoted in Greenblatt and Powell, "Rise of Megachurches," 778.

33. Greenblatt and Powell, "Rise of Megachurches," 778.

34. McFarland as quoted in Greenblatt and Powell, "Rise of Megachurches," 782.

35. Van Biema and Chu, "Does God Want You to Be Rich?"

36. Van Biema and Chu, "Does God Want You to Be Rich?"

CHAPTER 11: WEAPONS OF MASS CONSUMPTION

1. Gerri Hirshey, "Time to Buy a New Stove Again," *New York Times*, December 14, 2008, 4.

2. Joseph Guiltinan, "Creative Destruction and Destructive Creations: Environmental Ethics and Planned Obsolescence," *Journal of Business Ethics* 889 (2009): 19–28, 20.

3. Guiltinan, "Creative Destruction and Destructive Creations," 20.

4. K. A. Harmer, "Organized Waste: The History of Planned Obsolescence from the 1930s to the Present Day," presented at Waste: The Social Context Conference, May 11 to 14, Edmonton, Alberta, Canada, 2005, 258.

5. Harmer, "Organized Waste," 258.

6. Quoted in Nick Nettles, "Designing for Destruction," *Ecologist*, July 2008, 48.

7. Guiltinan, "Creative Destruction and Destructive Creations," 19.

8. Tim Cooper, "Inadequate Life? Evidence of Consumer Attitudes to Product Obsolescence," *Journal of Consumer Policy* 27 (2004): 421–49.

9. Susan Kretchmer, "Advertainment: The Evolution of Product Placement as a Mass Marketing Strategy," in *Handbook of Product Placement in the Mass Media*, ed. Mary-Lou Galician (Philadelphia: Taylor and Francis Group, 2004), 37–54.

10. Maria Esposito, "Will Advertainment Be Curbed in the States?" *The World*, July 11, 2008, 19.

11. CFR Title 47 Part 73.1212—Sponsorship Identification, www.Cfr.vlex.com.

12. Esposito, "Will Advertainment Be Curbed in the States?" 19.

13. Esposito, "Will Advertainment Be Curbed in the States?" 19.

14. Esposito, "Will Advertainment Be Curbed in the States?" 19.

15. Simon Hudson, David Hudson, and John Peloza, "Meet the Parents," *Journal of Business Ethics*, 80 (2008): 290–291.

16. Kretchmer, "Advertainment," 43.

17. Kathleen J. Turner, "Insinuating the Product into the Message: An Historical Context for Product Placement," in Galician, *Product Placement in the Mass Media*, 11.

18. Turner, "Insinuating the Product into the Message," 11.

19. Turner, "Insinuating the Product into the Message," 11.

20. Turner, "Insinuating the Product into the Message," 11.

21. Turner, "Insinuating the Product into the Message," 12.

22. Turner, "Insinuating the Product into the Message," 12.

23. Turner, "Insinuating the Product into the Message," 14.

24. Stephanie Mansfield, "Sweet Success: Reese's Cashes in on E.T.'s Candy Cravings," *Washington Post*, July 14, 1982.

25. "Brandnamed-Films," www.brandchannel.com/brandcameo_films.asp, accessed June 19, 2009.

26. "Brandnamed-Films."

27. Moonhee Yang and David R. Roskos-Ewoldsen, "The Effectiveness of Brand Placements in the Movies," *Journal of Communication* 57 (2007): 469–89.

28. Ravi Somaiya, "Media: Product Placement: Chloe, Is Jack Who Does Our Phones?" *The Guardian*, June 16, 2008, 3.

29. AC Nielsen Company, 2011.

30. Joyce Julius Associates, 2011.

31. Yang and Roskos-Ewoldsen, "The Effectiveness of Brand Placements in the Movies," 470–71.

32. Kretchmer, "Advertainment," 48.

33. Kretchmer, "Advertainment," 49–50.

34. Beth Snyder Bulik, "What Do Video Games Have to Do with Marketing?" *Advertising Age*, 80, no. 11, (March 30, 2009): 54.

35. Quotes from this and preceding paragraph come from Joy Lanzendorfer, "Is (More) Product Placement in Young-Adult Books Inevitable?" *Writer* 121, September 2008, 8–9.

36. Lanzendorfer, "Is (More) Product Placement"; Simon Hudson, David Hudson, and John Peloza, "Meet the Parents," 289.

37. Elizabeth Bell, *Theories of Performance* (Thousand Oaks, CA: Sage Publications, 2008), www.worldcat.org/oclc/166387386, accessed July 1, 2009.

38. Yong Liu, "Word of Mouth for Movies: Its Dynamics and Impact on Box Office Revenue," *Journal of Marketing* 70 (July 2006): 74–89.

39. Kristen L. Smith, "Six Ways to Leverage Word-of-Mouth Marketing," *Promo*, March 1, 2009, www.promomagazine.com/viralmarketing/03 01-wordofmouth-marketing-leverage, accessed June 8, 2009.

40. Nessim Hanna, Richard Wozniak, and Margaret Hanna, *Consumer Behavior*, 2nd ed., (Dubuque, IA: Kendall Hunt, 2006).

41. Rob Walker, "The Hidden (in Plain Sight) Persuaders," *New York Times*, December 5, 2004, www.nytimes.com, accessed July 3, 2009.

42. Laurie Burkitt, "Word-of-Mouth Marketing Evolution," www.forbes.com /2009/05/21/bzzagent-bzzscape-womm-leadership-cmo-network-bzz, accessed June 8, 2009.

43. www.aboutbzzagent.com/word-of-mouth/company/leadership, accessed July 21, 2009; Walker, "The Hidden (in Plain Sight) Persuaders."

44.   Walker, "The Hidden (in Plain Sight) Persuaders."

45.   www.womma.org, June 24, 2009.

46.   Smith, "Six Ways to Leverage Word-of-Mouth Marketing."

47.   Melanie Wells, "Kid Nabbing," *Forbes*, February 2, 2004, www.forbes .com/forbes/2004/02021084.html, accessed June 18, 2009; Philip Kotler and Gary Armstrong, *Principles of Marketing* (Upper Saddle River, NJ: Prentice Hall, 2009), 136–37; Robert Berner, "I Sold It Through the Grapevine," *Business Week*, May 29, 2006, 140.

48.   Berner, "I Sold It Through the Grapevine."

CHAPTER 12: THE THREE INGREDIENTS OF SELF-CONTROL

1.   Katherine Rosman, "Blackberry Orphans," *Wall Street Journal*, December 8, 2006, W1; Anjali Athavaley, "The New Blackberry Addicts," *Wall Street Journal*, January 23, 2007, D1.

2.   Ian McDonald, "Benefits of Saving Wasted on Youth," *Wall Street Journal*, August 7, 2006, R1; Eleanor Laise, "Employers Grab Reins of Workers' 401(k)s," *Wall Street Journal*, April 25, 2007, D1.

3.   Ben Stein, "Living Hand to Mouth—and Barely Getting By," *Yahoo Finance*, www.finance.yahoo.com/columnist/article/yourlife/24492p=1, accessed January 5, 2010.

4.   Roy F. Baumeister and Todd F. Heatherton, "Self-Regulation Failure: An Overview," *Psychological Inquiry* 7 (1996): 1–15.

5.   W. Mischel, Y. Shoda and P. K. Peake, "The Nature of Adolescent Competencies Predicted by Preschool Delay of Gratification," *Journal of Personality and Social Psychology* 54 (1988): 687–96.

6.   Mark Muraven, Dianne Tice, and Roy F. Baumeister, "Self-Control as Limited Resource: Regulatory Depletion Patterns," *Journal of Personality and Social Psychology* 74 (1998): 774–89.

7.   Roy F. Baumeister, Todd F. Heatherton, and Dianne M. Tice, *Losing Control: How and Why People Fail at Self-Regulation* (San Diego: Academic Press, 1994).

8.   Baumeister and Heatherton, "Self-Regulation Failure: An Overview," 1–15, 2.

9.   Roy F. Baumeister, "Yielding to Temptation: Self-Control Failure, Impulsive Purchasing, and Consumer Behavior," *Journal of Consumer Research* 28, no. 4 (2002): 670–76.

10.   Clayton E. Tucker-Ladd, *Psychological Self-Help* (downloadable online book), Self-Help Foundation, 1996–2004, www.psychologicalselfhelp.org, accessed November 17, 2009, 252.

11. Tucker-Ladd, *Psychological Self-Help*, 1097.

12. www.ipip.ori.org, accessed June 24, 2007.

13. Baumeister, "Yielding to Temptation," 670–76.

14. Baumeister, "Yielding to Temptation," 670–76; Burroughs and Rindfleisch, "Materialism and Well-Being," 348-370.

15. Dianne M. Tice, E. Bratslausky, and Roy F. Baumeister, "Emotional Distress Regulation Takes Precedence Over Impulse Control," *Journal of Personality and Social Behavior* 80 (2001): 53–67.

16. Burroughs and Rindfleisch, "Materialism and Well-Being."

17. James A. Roberts and Chris Manolis, "Cooking Up a Recipe for Self-Control," unpublished manuscript, 2009, 15–16.

18. Elizabeth Weise, "Idea of Simple Life Takes Hold," *USA Today*, March 22, 2006, www.usatoday.com/news/nation/2006-03-22-simple-life_x.htm, accessed March 24, 2006.

19. Weise, "Idea of Simple Life Takes Hold."

20. Lynn Schnurnberger, "Live Well with What You Have," *Parade*, January 10, 2010, 6.

21. Baumeister, "Yielding to Temptation," 670–76.

22. David Meyers, *Psychology*.

23. Meyers, 551.

24. Meyers, 554.

25. Baumeister, "Yielding to Temptation," 670–76.

26. Kathleen Vohs et al., "Making Choices Impairs Subsequent Self-Control," *Personality Processes and Individual Differences* 94, no. 5 (2008): 883–98.

27. C. Martijn et al., "Getting a Grip on Ourselves," *Social Cognition* 20, no. 6 (2002): 441–60.

28. Roberts and Manolis, "Cooking Up a Recipe for Self-Control," 15.

CHAPTER 13: STEP AWAY FROM THE SHOPPING CART

1. Douglas Belkin, "A Bank Run Teaches the 'Plain People' about the Risks of Modernity," *Wall Street Journal*, July 1, 2009, www.online.wsj.com/article/513124640811-360577075.html, accessed July 6, 2009.

2. Richard H. Thaler and Cass Sunstein, *Nudge: Improving Decisions About Health, Wealth, and Happiness* (London: Penguin Books, 2009), 114–17.

3. Dave Ramsey, Financial Peace University, www.daveramsey.com, accessed January 21, 2010.

4. Tucker-Ladd, *Psychological Self-Help*.

5. Thaler and Sunstein, *Nudge*, 718.

CHAPTER 14: THE CARROT AND THE STICK

1. Ramsey, Financial Peace University, www.daveramsey.com, accessed January 21, 2010.
2. Myers, *Psychology*, 323.
3. B. F. Skinner, Behavioral Learning Solutions, www.blsolutionsaba.org/aba, accessed January 24, 2010.
4. Tucker-Ladd, *Psychological Self-Help*.
5. Tucker-Ladd, *Psychological Self-Help*.
6. Tucker-Ladd, *Psychological Self-Help*.
7. Tucker-Ladd, *Psychological Self-Help*, 1143–44.
8. Tucker-Ladd, *Psychological Self-Help*, 1143–45.

CHAPTER 15: YOUR MONEY OR YOUR LIFE

1. Tucker-Ladd, *Psychological Self-Help*.
2. Harold Kushner, *When All You've Ever Wanted Isn't Enough* (New York: Pocket Books, 1986), 23.
3. Kushner, *When All You've Ever Wanted Isn't Enough*, 42–43.
4. www.quotegarden.com/life.html, accessed July 22, 2009.
5. www.thinkexist.com/quotes/aristotle, accessed January 5, 2010.
6. Carolinne White, *The Confessions of St. Augustine*, (Grand Rapids, MI: William B. Eerdmans Publishing Company, 2001), 28.
7. Quoted in Kent Swift, "Financial Success and the Good Life: What Have We Learned from Empirical Studies?" *Journal of Business Ethics* 75 (2007): 195.
8. Tucker-Ladd, *Psychological Self-Help*, 116.
9. Kushner, *When All You've Ever Wanted Isn't Enough*.
10. Kushner, *When All You've Ever Wanted Isn't Enough*, 29.
11. Tucker-Ladd, *Psychological Self-Help*, 140–41.
12. Tucker-Ladd, *Psychological Self-Help*, 113.
13. Lawrence Kohlberg, *Essays on Moral Development*, vol. 1, *The Philosophy of Moral Development* (San Francisco: Harper & Row, 1981), 1.
14. Kohlberg, *Essays on Moral Development*, Chapter 1.
15. Tucker-Ladd, *Psychological Self-Help*, 129.
16. www.worldofquotes.com/topic/materialism/index.html, accessed July 17, 2009.

17. Joan Smith, "Spend! Spend! Spend!," *The Times* (UK), October 2, 2003, 24.
18. De Graaf, Wann, and Naylor, *Affluenza*.
19. Kushner, *When All You've Ever Wanted Isn't Enough*, 18.
20. www.brainyquote.com/quotes/quotes/r/ralphwaldo103471.html, accessed January 15, 2010.

INDEX

Page numbers of charts and graphs appear in italics.

and, 90–93; Mental Illness Around
the World, *91–92*
Dichter, Ernest, 50
Diener, Ed, 84, 87, 155
Diffusion of Innovation (Rogers), 229
Dollar, Creflo, 191, 192, 193
Dollar, Taffi, 191
Dominos Pizza, 226
dreams, 176–77
Dunn, Elizabeth, 162

Easterlin, Richard, 83–84
Economic Mobility Project, 28
ego-depletion, 242–43, 254–55; decision-
making and, 257–58; scale, 258–60
Einstein, Albert, 45, 77
electronics, 12; addictive quality, 225,
239–40; discarded, 210, 212; environmental programming for consumers and, 262; perceived
obsolescence, 212; planned obsolescence, 207, 208–10
Emerson, Ralph Waldo, 239, 308
energy: Arab oil embargo, 1973, 56;
Carter's plan, 59–60; crises and
inflation, 56; Reagan's approach, 60
envelope strategy, 269, 272
environmental programming for consumers, 18–19, 261–80; five steps for
tweaking, 276–77; Save More
Tomorrow (SMT) program, 266–67,
277; twenty-five tweaks to financial
peace, 263–75; worksheet, 278–80
Epic of America, The (Adams), 37, 38
"escape theory," 107
E.T. (film), 219–20, 222
eToys, 65
Evan Almighty (film), 202
evolutionary psychology, 178–79
"extended self," 6, 105

F. W. Woolworth, 31
Fannie Mae and Freddie Mac, 69,
72, 74

Federal Housing Administration
(FHA), 72, 73, 74; Urban Loan
program, 68, 72–73
Federal National Mortgage Assn.,
67, 72
FedEx, 220, 221
Feinberg, Richard, 127–28
finances (personal and household);
bankruptcy, 13, *14*, 104, 118, 135,
291; behavioral program for, 281–
94; Ben Franklin's advice, 19–20;
compulsive buying and, 104, 105–
6; Conflicting Values Scale, 251–
53; emergency fund, 265–66;
environmental program for, 261–
80; gap hypothesis and, 88–90;
household income and happiness,
84; life satisfaction and, 88, 123–
26, *124*, 240; "Love, Money, and
Life Satisfaction," *89*; materialism
and, 3, 8, 86, 90, 93; problems,
Americans with, 8, 113, 116; questions to ask yourself, 9; savings, 24,
47, 50, 51, *113*, 113–14, 240–41;
self-control and, 240–60; "Student
Satisfaction with Life Domains"
and, *88*; twelve signs of compulsive
debting, 111–12; twenty-five
tweaks to financial peace, 263–75.
See also budgeting; consumer debt;
credit cards; savings, personal
Firm, The (film), 220
Fitzgerald, F. Scott, 34
Fitzmaurice, Julie, 123
Flouri, Eirini, 166
Ford, Henry, 33
Ford Motor Company, 33, 35–36, 203
Founding of New England, The
(Adams), 37
"four freedoms," 40–43
Fox, Richard, 46
Frankl, Viktor, 291
Franklin, Benjamin, 19–20, 171
Franklin, Robert, 194